Georgia Milestones Assessment System Subject Test
Mathematics Grade 5

Student Practice Workbook
+ Two Full-Length GMAS Math Tests

Math Notion
www.MathNotion.com

GMAS Subject Test Mathematics Grade 5

Georgia Milestones Assessment System Subject Test

Mathematics Grade 5

Published in the United State of America By

The Math Notion

Web: WWW.MathNotion.com

Email: info@Mathnotion.com

Copyright © 2021 by the Math Notion. All rights reserved. No part of this publication may be reproduced, stored in a retrieval system, or transmitted in any form or by any means, electronic, mechanical, photocopying, recording, scanning, or otherwise, except as permitted under Section 107 or 108 of the 1976 United States Copyright Ac, without permission of the author.

All inquiries should be addressed to the Math Notion.

ISBN: 978-1-63620-083-5

> GMAS Subject Test Mathematics Grade 5

The Math Notion

Michael Smith has been a math instructor for over a decade now. He launched the Math Notion. Since 2006, we have devoted our time to both teaching and developing exceptional math learning materials. As a test prep company, we have worked with thousands of students. We have used the feedback of our students to develop a unique study program that can be used by students to drastically improve their math scores fast and effectively. We have more than a thousand Math learning books including:

– **SAT Math Prep**

– **ACT Math Prep**

– **SSAT/ISEE Math Prep**

–**Mathematics Prep Grade 3 to 8**

– **Common Core Math Prep**

–**many Math Education Workbooks, Study Guides, Practice and Exercise Books**

As an experienced Math test preparation company, we have helped many students raise their standardized test scores—and attend the colleges of their dreams: We tutor online and in person, we teach students in large groups, and we provide training materials and textbooks through our website and through Amazon.

You can contact us via email at:

info@Mathnotion.com

GMAS Subject Test Mathematics Grade 5

Get the Targeted Practice You Need to Ace the GMAS Math Test!

Georgia Milestones Assessment System Subject Test Mathematics Grade 5 includes easy-to-follow instructions, helpful examples, and plenty of math practice problems to assist students to master each concept, brush up their problem-solving skills, and create confidence.

The GMAS math practice book provides numerous opportunities to evaluate basic skills along with abundant remediation and intervention activities. It is a skill that permits you to quickly master intricate information and produce better leads in less time.

Students can boost their test-taking skills by taking the book's two practice GMAS Math exams. All test questions answered and explained in detail.

Important Features of the 5th grade GMAS Math Book:

- A **complete review** of GMAS math test topics,
- Over 2,500 practice problems covering all topics tested,
- The most important concepts you need to know,
- Clear and concise, easy-to-follow sections,
- Well designed for enhanced learning and interest,
- Hands-on experience with all question types,
- **2 full-length practice tests** with detailed answer explanations,
- Cost-Effective Pricing,

Powerful math exercises to help you avoid traps and pacing yourself to beat the Georgia GMAS test. Students will gain valuable experience and raise their confidence by taking 5th grade math practice tests, learning about test structure, and gaining a deeper understanding of what is tested on the GMAS math grade 5. If ever there was a book to respond to the pressure to increase students' test scores, this is it.

GMAS Subject Test Mathematics Grade 5

WWW.MathNotion.COM

… So Much More Online!

- ✓ FREE Math Lessons

- ✓ More Math Learning Books!

- ✓ Mathematics Worksheets

- ✓ Online Math Tutors

For a PDF Version of This Book

Please Visit WWW.MathNotion.com

GMAS Subject Test Mathematics Grade 5

Contents

Chapter 1 : Place Values and Number Sense 11
    Place Values 12
    Comparing and Ordering Numbers 13
    Numbers in Word Form 14
    Roman Numerals 15
    Rounding Numbers 16
    Odd or Even 17
    Repeating Patterns 18
    Growing Patterns 19
    Patterns: Numbers 20
    Answers of Worksheets 21

Chapter 2 : Whole Number Operations 23
    Adding Whole Numbers 24
    Subtracting Whole Numbers 25
    Multiplying Whole Numbers 26
    Dividing Hundreds 27
    Long Division by Two Digits 28
    Division with Remainders 28
    Rounding Whole Numbers 29
    Whole Number Estimation 30
    Answers of Worksheets 31

Chapter 3 : Number Theory 33
    Factoring Numbers 34
    Prime Factorization 34
    Divisibility Rules 35
    Greatest Common Factor 36
    Least Common Multiple 36
    Answers of Worksheets 37

Chapter 4 : Fractions and Mixed Numbers 39
    Simplifying Fractions 40
    Like Denominators 41
    Compare Fractions with Like Denominators 43
    More Than Two Fractions with Like Denominators 44
    Unlike Denominators 45
    Ordering Fractions 47

GMAS Subject Test Mathematics Grade 5

    Denominators of 10, 100, and 1000 48
    Fractions to Mixed Numbers 50
    Mixed Numbers to Fractions 51
    Add and Subtract Mixed Numbers 52
    Answers of Worksheets 53

Chapter 5 : Decimals **57**
    Adding and Subtracting Decimals 58
    Multiplying and Dividing Decimals 59
    Rounding Decimals 60
    Comparing Decimals 61
    Answers of Worksheets 62

Chapter 6 : Ratios and Rates **63**
    Simplifying Ratios 64
    Writing Ratios 64
    Create a Proportion 65
    Proportional Ratios 65
    Similar Figures 66
    Word Problems 67
    Answers of Worksheets 69

Chapter 7 : Measurement **71**
    Reference Measurement Units 72
    Metric Length Units 73
    Customary Length Units 73
    Metric Capacity Units 74
    Customary Capacity Units 74
    Metric Weight and Mass Units 75
    Customary Weight and Mass Units 75
    Temperature Units 76
    Time 77
    Money Amounts 78
    Money: Word Problems 79
    Answers of Worksheets 80

Chapter 8 : Algebraic Thinking **82**
    Finding Rules 83
    Algebraic Word Problems 84
    Evaluate Expressions 85
    Variables and Expressions 86
    Answers of Worksheets 87

GMAS Subject Test Mathematics Grade 5

Chapter 9 : Geometric 89
Identifying Angles 90
Estimate Angle Measurements 91
Measure Angles with a Protractor 92
Polygon Names 93
Classify Triangles 94
Parallel Sides in Quadrilaterals 95
Identify Rectangles 96
Perimeter: Find the Missing Side Lengths 97
Perimeter and Area of Squares 98
Perimeter and Area of rectangles 99
Find the Area or Missing Side Length of a Rectangle 100
Area and Perimeter: Word Problems 101
Circumference, Diameter, and Radius 102
Volume of Cubes and Rectangle Prisms 103
Answers of Worksheets 104

Chapter 10 : Three-Dimensional Figures 106
Identify Three-Dimensional Figures 107
Count Vertices, Edges, and Faces 108
Identify Faces of Three-Dimensional Figures 109
Answers of Worksheets 110

Chapter 11 : Symmetry and Transformations 111
Line Segments 112
Identify Lines of Symmetry 113
Count Lines of Symmetry 114
Parallel, Perpendicular and Intersecting Lines 115
Answers of Worksheets 116

Chapter 12 : Data Graphs, and Statistics 117
Mean and Median 118
Mode and Range 119
Graph Points on a Coordinate Plane 120
Bar Graph 121
Tally and Pictographs 122
Dot plots 123
Line Graphs 124
Stem-And-Leaf Plot 125
Scatter Plots 126
Probability Problems 127

GMAS Subject Test Mathematics Grade 5

Answers of Worksheets 128
Chapter 13 : GMAS Math Practice Tests 131
GMAS GRADE 5 MAHEMATICS REFRENCE MATERIALS 133
Georgia Milestones Assessment System Practice Test 1 135
Session 1 136
Session 2 140
Georgia Milestones Assessment System Practice Test 2 145
Session 1 146
Session 2 150
Chapter 14 : Answers and Explanations 155
Answer Key 155
Practice Test 1 157
Practice Test 2 161

GMAS Subject Test Mathematics Grade 5

Chapter 1 : Place Values and Number Sense

Topics that you'll learn in this chapter:

- ✓ Place Values,
- ✓ Compare and Ordering Numbers,
- ✓ Numbers in Word Form,
- ✓ Roman Numerals,
- ✓ Rounding Numbers,
- ✓ Odd or Even,
- ✓ Repeating Patterns,
- ✓ Growing Patterns,
- ✓ Patterns: Numbers,

GMAS Subject Test Mathematics Grade 5

Place Values

✍ Write numbers in expanded form.

1) Sixty–two   ___ + ___

2) fifty–six   ___ + ___

3) thirty–one   ___ + ___

4) forty–five   ___ + ___

5) twenty–eight   ___ + ___

✍ Circle the correct choice.

6) The 6 in 56 is in the

        Ones place   tens place   hundreds place

7) The 2 in 25 is in the

        Ones place   tens place   hundreds place

8) The 9 in 918 is in the

        Ones place   tens place   hundreds place

9) The 3 in 537 is in the

        Ones place   tens place   hundreds place

10) The 9 in 289 is in the

        Ones place   tens place   hundreds place

GMAS Subject Test Mathematics Grade 5

Comparing and Ordering Numbers

🖊 Use less than, equal to or greater than.

1) 31 _____ 33

2) 57 _____ 49

3) 92 _____ 88

4) 76 _____ 67

5) 43 _____ 43

6) 54 _____ 46

7) 97 _____ 88

8) 42 _____ 36

9) 55 _____ 55

10) 57 _____ 75

11) 28 _____ 38

12) 19 _____ 15

13) 82 _____ 90

14) 78 _____ 84

🖊 Order each set numbers from least to greatest.

15) – 18, – 22, 28, – 17, 4   ___, ___, ___, ___, ___, ___

16) 19, –36, 11, – 12, 5   ___, ___, ___, ___, ___, ___

17) 27, – 56, 20, 1, – 27   ___, ___, ___, ___, ___, ___

18) 26, – 96, 2, – 26, 87, –75   ___, ___, ___, ___, ___, ___

19) –10, –71, 70, –26, –59, –39   ___, ___, ___, ___, ___, ___

20) 88, 4, 38, 7, 78, 9   ___, ___, ___, ___, ___, ___

21) 84, 14, 24, 0, 35, 22   ___, ___, ___, ___, ___, ___

WWW.MathNotion.Com

GMAS Subject Test Mathematics Grade 5

Numbers in Word Form

✎ Write each number in words.

1) 372 _____

2) 605 _____

3) 550 _____

4) 351 _____

5) 793 _____

6) 647 _____

7) 3,219 _____

8) 5,326 _____

9) 2,842 _____

10) 4,691 _____

11) 5,531 _____

12) 7,360 _____

13) 2,532 _____

14) 8,014 _____

15) 11,242 _____

Roman Numerals

✎ Write in Romans numerals.

1	I	11	XI	21	XXI
2	II	12	XII	22	XXII
3	III	13	XIII	23	XXIII
4	IV	14	XIV	24	XXIV
5	V	15	XV	25	XXV
6	VI	16	XVI	26	XXVI
7	VII	17	XVII	27	XXVII
8	VIII	18	XVIII	28	XXVIII
9	IX	19	XIX	29	XXIX
10	X	20	XX	30	XXX

1) 11 _____   2) 21 _____

3) 24 _____   4) 16 _____

5) 27 _____   6) 29 _____

7) 12 _____   8) 28 _____

9) 15 _____   10) 20 _____

11) Add 16 + 14 and write in Roman numerals. _____

12) Subtract 34 – 5 and write in Roman numerals. _____

GMAS Subject Test Mathematics Grade 5

Rounding Numbers

🖎 Round each number to the underlined place value.

1) 3,793

2) 3,876

3) 3,452

4) 7,193

5) 5,278

6) 1,477

7) 8,313

8) 24.68

9) 84.92

10) 71.34

11) 664.7

12) 9,135

13) 15.381

14) 4,521

15) 36.50

16) 4,819

17) 6,685

18) 2,538

19) 73.62

20) 16,527

21) 29.720

22) 12,366

23) 31,729

24) 7,838

Odd or Even

☒ Identify whether each number is even or odd.

1) 18 _____

2) 27 _____

3) 21 _____

4) 17 _____

5) 67 _____

6) 76 _____

7) 80 _____

8) 53 _____

9) 58 _____

10) 98 _____

11) 49 _____

12) 113 _____

☒ Circle the even number in each group.

13) 52, 11, 35, 73, 5, 29

14) 13, 15, 113, 87, 71, 18

15) 33, 45, 86, 59, 63, 87

16) 55, 32, 79, 51, 21, 83

☒ Circle the odd number in each group.

17) 54, 36, 48, 76, 71, 100

18) 32, 56, 40, 74, 98, 67

19) 58, 92, 25, 78, 76, 50

20) 89, 12, 88, 42, 48, 120

GMAS Subject Test Mathematics Grade 5

Repeating Patterns

✎ Circle the picture that comes next in each picture pattern.

1)

 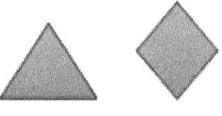

2)

3)

 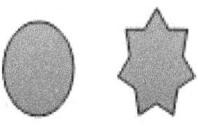

4)

5)

Growing Patterns

✎ Draw the picture that comes next in each growing pattern.

1)

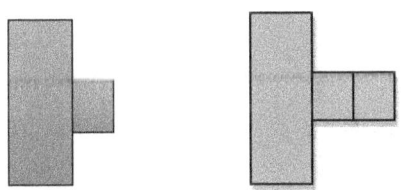

2)

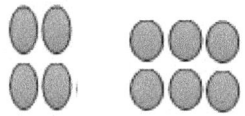

3)

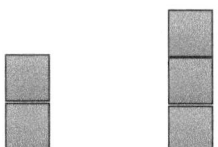

4)

5)

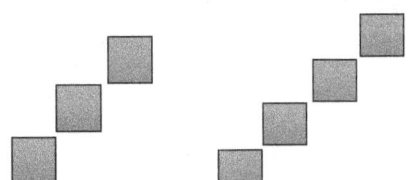

GMAS Subject Test Mathematics Grade 5

Patterns: Numbers

✎ Write the numbers that come next.

1) 2, 5, 8, 11,   ____, ____, ____, ____

2) 10, 15, 20, 25,   ____, ____, ____, ____

3) 4, 8, 12, 16,   ____, ____, ____, ____

4) 7, 17, 27, 37,   ____, ____, ____, ____

5) 5, 12, 19, 26,   ____, ____, ____, ____

6) 8, 16, 24, 32, 40,   ____, ____, ____, ____

✎ Write the next three numbers in each counting sequence.

1) −31, −19, −7, _____, _____, _____, _____

2) 541, 526, 511, _____, _____, _____, _____

3) 14, 34, _____, _____, 94, _____

4) 21, 29, _____, _____, _____

5) 89, 78, _____, _____, _____

6) 95, 82, 69, _____, _____, _____

7) 198, 166, 134, _____, _____, _____

8) What are the next three numbers in this counting sequence?

   1870, 1970, 2070, _____, _____, _____

9) What is the fourth number in this counting sequence?

   8, 14, 20, _____

GMAS Subject Test Mathematics Grade 5

Answers of Worksheets

Place Values

1) 60 + 2
2) 50 + 6
3) 30 + 1
4) 40 + 5
5) 20 + 8
6) ones place
7) tens place
8) hundreds place
9) tens place
10) ones place

Comparing and Ordering Numbers

1) 31 less than 33
2) 57 greater than 49
3) 92 greater than 88
4) 76 greater than 67
5) 43 equals to 43
6) 54 greater than 46
7) 97 greater than 88
8) 42 greater than 36
9) 55 equals to 55
10) 57 less than 75
11) 28 less than 38
12) 19 greater than 15
13) 82 less than 90
14) 78 less than 84
15) −22, −18, −17, 4, 28
16) −36, −12, 5, 11, 19
17) −56, −27, 1, 20, 27
18) −96, −75, −26, 2, 26, 87
19) −71, −59, −39, −26, −10, 70
20) 4, 7, 9, 38, 78, 88
21) 0, 14, 22, 24, 35, 84

Numbers in Word Form

1) three hundred seventy-two
2) six hundred five
3) five hundred fifty
4) three hundred fifty-one
5) seven hundred ninety-three
6) six hundred forty-seven
7) three thousand, two hundred nineteen
8) five thousand, three hundred twenty-six
9) two thousand, eight hundred forty-two
10) four thousand, six hundred ninety-one
11) five thousand, five hundred thirty-one
12) seven thousand, three hundred sixty
13) two thousand, five hundred thirty-two
14) eight thousand, fourteen
15) eleven thousand, two hundred forty-two

Roman Numerals

1) XI
2) XXI
3) XXIV
4) XVI
5) XXVII
6) XXIX
7) XII
8) XXVIII
9) XV
10) XX
11) XXX
12) XXIX

Rounding Numbers

1) 4,000
2) 4,000
3) 3,450
4) 7,190
5) 5,280
6) 1,480
7) 8,300
8) 24.70
9) 85.00
10) 71.30
11) 665.00
12) 9,000

WWW.MathNotion.Com

GMAS Subject Test Mathematics Grade 5

13) 15.380	16) 4,800	19) 73.60	22) 12,370
14) 4,500	17) 6,700	20) 16,500	23) 31,700
15) 37.00	18) 2,540	21) 30.00	24) 7,840

Odd or Even

1) even	6) even	11) odd	16) 32
2) odd	7) even	12) odd	17) 71
3) odd	8) odd	13) 52	18) 67
4) odd	9) even	14) 18	19) 25
5) odd	10) even	15) 86	20) 89

Repeating pattern

1)  2)  3)

4)  5)

Growing patterns

1) 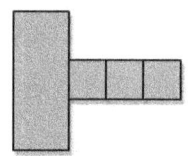 2) 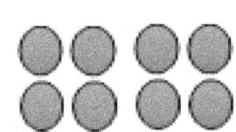 3)

4) 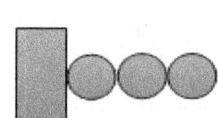 5)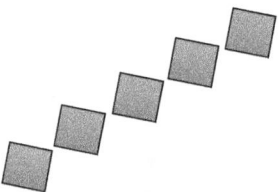

Patterns: Numbers

1) 2, 5, 8, 11, 14, 17, 20, 23   4) 7, 17, 27, 37, 47, 57, 67, 77

2) 10, 15, 20, 25, 30, 35, 40, 45   5) 5, 12, 19, 26, 33, 40, 47, 54

3) 4, 8, 12, 16, 20, 24, 28, 32   6) 8, 16, 24, 32, 40, 48, 56, 64

Patterns

1) 5, 17, 29, 41   4) 37, 45, 53   7) 102, 70, 38

2) 496, 481, 466, 451   5) 67, 56, 45   8) 2170, 2270, 2370

3) 14, 34, 54, 74, 94, 114 6) 56, 43, 30   9) 26

GMAS Subject Test Mathematics Grade 5

Chapter 2 : Whole Number Operations

Topics that you'll learn in this chapter:

- ✓ Adding Whole Numbers,
- ✓ Subtracting Whole Numbers,
- ✓ Multiplying Whole Numbers,
- ✓ Dividing Hundreds,
- ✓ Long Division by One Digit,
- ✓ Division with Remainders,
- ✓ Rounding Whole Numbers,
- ✓ Whole Number Estimation,

GMAS Subject Test Mathematics Grade 5

Adding Whole Numbers

✏ Add.

1) 5,763
   + 8,238
   _____

2) 6,834
   + 4,998
   _____

3) 3,548
   + 5,693
   _____

4) 2,769
   + 8,872
   _____

5) 3,196
   + 2,936
   _____

6) 7,009
   + 4,992
   _____

✏ Find the missing numbers.

7) 3,468 + ___ = 4,102

8) 840 + 2,360 = ___

9) 5,200 + ___ = 7,980

10) 631 + ___ = 2,007

11) ___ + 803 = 3,945

12) ___ + 2,156 = 5,922

13) David sells gems. He finds a diamond in Istanbul and buys it for $4,795. Then, he flies to Cairo and purchases a bigger diamond for the bargain price of $9,633. How much does David spend on the two diamonds? _____

WWW.MathNotion.Com

Subtracting Whole Numbers

✎ Subtract.

1) 10,512 − 4,411 = _____

2) 5,204 − 3,679 = _____

3) 8,520 − 6,483 = _____

4) 8,001 − 5,224 = _____

5) 11,916 − 8,711 = _____

6) 5,005 − 2,008 = _____

✎ Find the missing number.

7) 5,263 − ___ = 2,367

8) 7,198 − ___ = 4,742

9) 8,928 − 3,764 = ___

10) 6,511 − ___ = 3,759

11) 7,003 − 5,489 = ___

12) 8,800 − 5,995 = ___

13) Jackson had $7,189 invested in the stock market until he lost $3,793 on those investments. How much money does he have in the stock market now?

_____

GMAS Subject Test Mathematics Grade 5

Multiplying Whole Numbers

✏️ Find the answers.

1) 2,200 × 31

2) 3,200 × 22

3) 5,790 × 5

4) 5,220 × 3

5) 6,911 × 3

6) 1,998 × 40

7) 2,893 × 5.5

8) 2,254 × 3.5

9) 4,372 × 4.8

10) 3,984 × 2.75

11) 4,900 × 2.5

12) 8,200 × 4.5

Dividing Hundreds

📖 Find answers.

1) 4,440 ÷ 400

2) 1,600 ÷ 40

3) 9,990 ÷ 90

4) 4,200 ÷ 60

5) 6,400 ÷ 8,000

6) 2,700 ÷ 30

7) 3,333 ÷ 30

8) 558 ÷ 45

9) 2,278 ÷ 85

10) 1,683 ÷ 55

11) 1,582 ÷ 35

12) 9,000 ÷ 600

13) 1,000 ÷ 2,500

14) 44.8 ÷ 20

15) 6,800 ÷ 400

16) 1,500 ÷ 5,000

17) 36.60 ÷ 120

18) 7,700 ÷ 700

19) 5,400 ÷ 600

20) 8,000 ÷ 160

21) 18,000 ÷ 9,000

22) 42,000 ÷ 30

23) 480 ÷ 40

24) 63,000 ÷ 900

GMAS Subject Test Mathematics Grade 5

Long Division by Two Digits

✏ Find the quotient.

1) 18)576

2) 14)952

3) 21)588

4) 23)299

5) 44)748

6) 26)234

7) 16)496

8) 29)1,479

9) 54)1,080

10) 41)1,476

11) 53)2,491

12) 60)2,880

13) 32)2,912

14) 77)8,393

15) 85)3,740

16) 57)4,617

17) 50)9,200

18) 25)15,400

Division with Remainders

✏ Find the quotient with remainder.

1) 14)715

2) 16)2,750

3) 27)4,603

4) 58)2,554

5) 42)7,732

6) 63)6,737

7) 71)9,036

8) 65)8,624

9) 35)5,705

10) 92)13,161

11) 46)12,214

12) 69)42,482

13) 85)6,858

14) 87)34,304

WWW.MathNotion.Com

GMAS Subject Test Mathematics Grade 5

Rounding Whole Numbers

Round each number to the underlined place value.

1) 7,<u>5</u>33
2) 9,<u>3</u>74
3) 8,8<u>8</u>3
4) 2,3<u>6</u>8
5) 5,5<u>7</u>7
6) 3,3<u>8</u>1
7) 3,<u>5</u>20
8) 9,3<u>3</u>8
9) 8.<u>5</u>81

10) 33.<u>5</u>7
11) 51.6<u>9</u>
12) 22.<u>1</u>38
13) <u>6</u>,758
14) 11,5<u>5</u>7
15) 8,8<u>3</u>8
16) 5.8<u>8</u>9
17) 1.8<u>6</u>0
18) 25.<u>0</u>70

19) <u>9</u>.332
20) 49.4<u>8</u>
21) 28.<u>8</u>9
22) 24,3<u>7</u>7
23) 52,1<u>5</u>8
24) 13,8<u>8</u>3
25) 9,<u>6</u>09
26) 17,4<u>5</u>1
27) 18,<u>7</u>68

WWW.MathNotion.Com

GMAS Subject Test Mathematics Grade 5

Whole Number Estimation

✎ Estimate the sum by rounding each added to the nearest ten.

1) 875 + 325

2) 985 + 1,452

3) 2,424 + 4,128

4) 1,576 + 6,279

5) 1,247 + 3,863

6) 6,746 + 5,121

7) 3,924 + 6,456

8) 1,785 + 7,164

9) 1,458
   + 2,442
   ─────

10) 5,689
    + 4,151
    ─────

11) 8,259
    + 4,754
    ─────

12) 6,788
    + 3,954
    ─────

13) 9,123
    + 4,455
    ─────

14) 6,680
    + 5,358
    ─────

15) 3,165
    + 7,124
    ─────

16) 8,859
    + 6,452
    ─────

GMAS Subject Test Mathematics Grade 5

Answers of Worksheets

Adding Whole Numbers

1) 14,001
2) 11,832
3) 9,241
4) 11,641
5) 6,132
6) 12,001
7) 634
8) 3,200
9) 2,780
10) 1,376
11) 3,142
12) 3,766
13) $14,428

Subtracting Whole Numbers

1) 6,101
2) 1,525
3) 2,037
4) 2,777
5) 3,205
6) 2,997
7) 2,896
8) 2,456
9) 5,164
10) 2,752
11) 1,514
12) 2,805
13) 3,396

Multiplying Whole Numbers

1) 68,200
2) 70,400
3) 28,950
4) 15,660
5) 20,733
6) 79,920
7) 15,911.5
8) 7,889
9) 20,985.6
10) 10,956
11) 12,250
12) 36,900

Dividing Hundreds

1) 11.1
2) 40
3) 111
4) 70
5) 0.8
6) 90
7) 111.1
8) 12.4
9) 26.8
10) 30.6
11) 45.2
12) 15
13) 0.4
14) 2.24
15) 17
16) 0.3
17) 0.305
18) 11
19) 9
20) 50
21) 2
22) 1,400
23) 12
24) 70

Long Division by Two Digits

1) 32
2) 68
3) 28
4) 13
5) 17
6) 9
7) 31
8) 51
9) 20
10) 36
11) 47
12) 48
13) 91
14) 109
15) 44
16) 81
17) 184
18) 616

GMAS Subject Test Mathematics Grade 5

Division with Remainders

1) 51 R1
2) 171 R14
3) 170 R13
4) 44 R2
5) 184 R4
6) 106 R59
7) 127 R19
8) 132 R44
9) 163 R0
10) 143 R5
11) 265 R24
12) 615 R47
13) 80 R58
14) 394 R26

Rounding Whole Numbers

1) 7,500
2) 9,400
3) 8,880
4) 2,370
5) 5,580
6) 3,380
7) 3,500
8) 9,340
9) 8.60
10) 33.60
11) 51.70
12) 22.100
13) 7,000
14) 11,560
15) 8,840
16) 5.900
17) 1.900
18) 25.100
19) 9.000
20) 49.50
21) 28.90
22) 24,380
23) 52,160
24) 13,880
25) 9,600
26) 17,450
27) 18,800

Whole Number Estimation

1) 1,200
2) 2,440
3) 6,550
4) 7,860
5) 5,110
6) 11,870
7) 10,380
8) 8,950
9) 3,900
10) 9,840
11) 13,010
12) 10,740
13) 13,580
14) 12,040
15) 10,290
16) 15,310

Chapter 3 : Number Theory

Topics that you'll learn in this chapter:

- ✓ Factoring Numbers,
- ✓ Prime Factorization,
- ✓ Divisibility Rules,
- ✓ Greatest Common Factor,
- ✓ Least Common Multiple,

GMAS Subject Test Mathematics Grade 5

Factoring Numbers

✏ List all positive factors of each number.

1) 12	6) 56	11) 27
2) 16	7) 65	12) 63
3) 28	8) 70	13) 72
4) 34	9) 25	14) 15
5) 95	10) 48	15) 80

✏ List the prime factorization for each number.

16) 10	19) 30	22) 55
17) 26	20) 40	23) 78
18) 20	21) 44	24) 96

Prime Factorization

✏ Factor the following numbers to their prime factors.

1) 6	9) 58	17) 69
2) 49	10) 62	18) 76
3) 60	11) 75	19) 86
4) 4	12) 88	20) 92
5) 46	13) 93	21) 99
6) 57	14) 100	22) 77
7) 54	15) 68	23) 90
8) 38	16) 90	24) 74

GMAS Subject Test Mathematics Grade 5

Divisibility Rules

✎ Use the divisibility rules to underline the factors of the number.

1) 8   2 3 4 5 6 7 8 9 10

2) 18   2 3 4 5 6 7 8 9 10

3) 55   2 3 4 5 6 7 8 9 10

4) 45   2 3 4 5 6 7 8 9 10

5) 20   2 3 4 5 6 7 8 9 10

6) 9   2 3 4 5 6 7 8 9 10

7) 21   2 3 4 5 6 7 8 9 10

8) 28   2 3 4 5 6 7 8 9 10

9) 36   2 3 4 5 6 7 8 9 10

10) 40 2 3 4 5 6 7 8 9 10

11) 39 2 3 4 5 6 7 8 9 10

12) 51 2 3 4 5 6 7 8 9 10

GMAS Subject Test Mathematics Grade 5

Greatest Common Factor

✎ Find the GCF for each number pair.

1) 25, 15	9) 52, 3	17) 66, 18
2) 8, 18	10) 12, 54	18) 70, 15
3) 14, 28	11) 11, 13	19) 38, 14
4) 18, 32	12) 56, 48	20) 36, 28
5) 15, 45	13) 75, 25	21) 100, 60
6) 22, 33	14) 40, 60	22) 85, 35
7) 19, 21	15) 52, 32	23) 16, 48
8) 27, 72	16) 30, 55	24) 13, 39

Least Common Multiple

✎ Find the LCM for each number pair.

1) 3, 15	9) 13, 26	17) 13, 2, 26
2) 5, 35	10) 15, 65	18) 18, 6, 24
3) 24, 16	11) 12, 8	19) 9, 12, 15
4) 28, 40	12) 6, 44	20) 7, 12, 4
5) 9, 27	13) 10, 16	21) 5, 15, 16
6) 46, 23	14) 7, 6	22) 13, 4, 26
7) 22, 66	15) 12, 36, 24	23) 3, 14, 5
8) 4, 9	16) 5, 11, 2	24) 32, 8, 3

WWW.MathNotion.Com

GMAS Subject Test Mathematics Grade 5

Answers of Worksheets

Factoring Numbers

1) 1, 2, 3, 4, 6, 12
2) 1, 2, 4, 8, 16
3) 1, 2, 4, 7, 14, 28
4) 1, 2, 17, 34
5) 1, 5, 19, 95
6) 1, 2, 4, 7, 8, 14, 28, 56
7) 1, 5, 13, 65
8) 1, 2, 5, 7, 10, 14, 35, 70
9) 1, 5, 25
10) 1, 2, 3, 4, 6, 8, 12, 16, 24, 48
11) 1, 3, 9, 27
12) 1, 3, 7, 9, 21, 63
13) 1, 2, 3, 4, 6, 8, 9, 12, 18, 24, 36, 72
14) 1, 3, 5, 15
15) 1, 2, 4, 5, 8, 10, 16, 20, 40, 80
16) 2 × 5
17) 2 × 13
18) 2 × 2 × 5
19) 2 × 3 × 5
20) 2 × 2 × 2 × 5
21) 2 × 2 × 11
22) 5 × 11
23) 2 × 3 × 13
24) 2 × 2 × 2 × 2 × 2 × 3

Prime Factorization

1) 2. 3
2) 7. 7
3) 2. 2. 3. 5
4) 2. 2
5) 2. 23
6) 3. 19
7) 2. 3. 3. 3
8) 2. 19
9) 2. 29
10) 2. 31
11) 3. 5. 5
12) 2. 2. 2. 11
13) 3. 31
14) 2. 2. 5. 5
15) 2. 2. 17
16) 2. 3. 3. 5
17) 3. 23
18) 2. 2. 19
19) 2. 43
20) 2. 2. 23
21) 3. 3. 11
22) 7. 11
23) 2. 3. 3. 5
24) 2. 37

Divisibility Rules

1) 8   <u>2</u> 3 <u>4</u> 5 6 7 <u>8</u> 9 10
2) 18   <u>2</u> <u>3</u> 4 5 <u>6</u> 7 8 <u>9</u> 10
3) 55   2 3 4 <u>5</u> 6 7 8 9 10
4) 45   2 <u>3</u> 4 <u>5</u> 6 7 8 <u>9</u> 10
5) 20   <u>2</u> 3 <u>4</u> <u>5</u> 6 7 8 9 <u>10</u>
6) 9   2 <u>3</u> 4 5 6 7 8 <u>9</u> 10

GMAS Subject Test Mathematics Grade 5

7) 21   2 <u>3</u> 4 5 6 <u>7</u> 8 9 10
8) 28   <u>2</u> 3 <u>4</u> 5 6 <u>7</u> 8 9 10
9) 36   <u>2</u> <u>3</u> <u>4</u> 5 <u>6</u> 7 8 <u>9</u> 10
10) 40   <u>2</u> 3 <u>4</u> <u>5</u> 6 7 <u>8</u> 9 <u>10</u>
11) 39   2 <u>3</u> 4 5 6 7 8 9 10
12) 51   2 <u>3</u> 4 5 6 7 8 9 10

Greatest Common Factor

1) 5   7) 1   13) 25   19) 2
2) 2   8) 9   14) 20   20) 4
3) 14   9) 1   15) 4   21) 20
4) 2   10) 6   16) 5   22) 5
5) 15   11) 1   17) 6   23) 16
6) 11   12) 8   18) 5   24) 13

Least Common Multiple

1) 15   7) 66   13) 80   19) 180
2) 35   8) 36   14) 42   20) 84
3) 48   9) 26   15) 72   21) 240
4) 280   10) 195   16) 110   22) 52
5) 27   11) 24   17) 26   23) 210
6) 46   12) 132   18) 72   24) 96

Chapter 4 : Fractions and Mixed Numbers

Topics that you'll learn in this chapter:

- ✓ Simplifying Fractions,
- ✓ Like Denominators,
- ✓ Compare Fractions with Like Denominators,
- ✓ More than two Fractions with Like Denominators,
- ✓ Unlike Denominators,
- ✓ Ordering Fractions,
- ✓ Denominators of 10, 100, and 1000,
- ✓ Fractions to Mixed Numbers,
- ✓ Mixed Numbers to Fractions,
- ✓ Add and Subtract Mixed Numbers,

GMAS Subject Test Mathematics Grade 5

Simplifying Fractions

✎ Simplify the fractions.

1) $\dfrac{44}{84}$

2) $\dfrac{8}{20}$

3) $\dfrac{12}{16}$

4) $\dfrac{4}{24}$

5) $\dfrac{15}{30}$

6) $\dfrac{9}{63}$

7) $\dfrac{4}{14}$

8) $\dfrac{17}{51}$

9) $\dfrac{24}{30}$

10) $\dfrac{5}{35}$

11) $\dfrac{16}{48}$

12) $\dfrac{33}{22}$

13) $\dfrac{45}{63}$

14) $\dfrac{2.4}{3.2}$

15) $\dfrac{12}{60}$

16) $\dfrac{70}{112}$

17) $\dfrac{2.7}{7.2}$

18) $\dfrac{33}{88}$

19) $\dfrac{1.5}{13.5}$

20) $\dfrac{39}{52}$

21) $\dfrac{5}{45}$

22) $\dfrac{2.1}{4.2}$

GMAS Subject Test Mathematics Grade 5

Like Denominators

✎ Add fractions.

1) $\dfrac{3}{4}+\dfrac{1}{4}$

2) $\dfrac{1}{5}+\dfrac{4}{5}$

3) $\dfrac{4}{9}+\dfrac{7}{9}$

4) $\dfrac{2}{7}+\dfrac{2}{7}$

5) $\dfrac{5}{13}+\dfrac{2}{13}$

6) $\dfrac{1}{14}+\dfrac{4}{14}$

7) $\dfrac{11}{19}+\dfrac{1}{19}$

8) $\dfrac{3}{16}+\dfrac{9}{16}$

9) $\dfrac{3}{10}+\dfrac{1}{10}$

10) $\dfrac{6}{17}+\dfrac{2}{17}$

11) $\dfrac{5}{22}+\dfrac{5}{22}$

12) $\dfrac{7}{35}+\dfrac{11}{35}$

13) $\dfrac{7}{27}+\dfrac{20}{27}$

14) $\dfrac{2}{31}+\dfrac{10}{31}$

15) $\dfrac{5}{23}+\dfrac{3}{23}$

16) $\dfrac{8}{41}+\dfrac{13}{41}$

17) $\dfrac{15}{37}+\dfrac{18}{37}$

18) $\dfrac{2}{51}+\dfrac{7}{51}$

19) $\dfrac{17}{26}+\dfrac{6}{26}$

20) $\dfrac{12}{48}+\dfrac{11}{48}$

21) $\dfrac{11}{29}+\dfrac{8}{29}$

22) $\dfrac{15}{34}+\dfrac{19}{34}$

23) $\dfrac{1}{19}+\dfrac{5}{19}$

24) $\dfrac{3}{53}+\dfrac{4}{53}$

25) $\dfrac{3}{20}+\dfrac{6}{20}$

26) $\dfrac{2}{63}+\dfrac{6}{63}$

27) $\dfrac{6}{38}+\dfrac{1}{38}$

28) $\dfrac{14}{31}+\dfrac{17}{31}$

29) $\dfrac{3}{28}+\dfrac{5}{28}$

30) $\dfrac{2}{37}+\dfrac{15}{37}$

WWW.MathNotion.Com

GMAS Subject Test Mathematics Grade 5

✎ **Subtract fractions.**

1) $\dfrac{8}{9} - \dfrac{4}{9}$

2) $\dfrac{3}{8} - \dfrac{1}{8}$

3) $\dfrac{9}{11} - \dfrac{3}{11}$

4) $\dfrac{9}{14} - \dfrac{4}{14}$

5) $\dfrac{15}{20} - \dfrac{8}{20}$

6) $\dfrac{8}{15} - \dfrac{7}{15}$

7) $\dfrac{11}{19} - \dfrac{9}{19}$

8) $\dfrac{13}{16} - \dfrac{1}{16}$

9) $\dfrac{7}{29} - \dfrac{4}{29}$

10) $\dfrac{14}{23} - \dfrac{7}{23}$

11) $\dfrac{15}{34} - \dfrac{7}{34}$

12) $\dfrac{18}{41} - \dfrac{9}{41}$

13) $\dfrac{17}{39} - \dfrac{16}{39}$

14) $\dfrac{6}{26} - \dfrac{2}{26}$

15) $\dfrac{14}{17} - \dfrac{4}{17}$

16) $\dfrac{33}{55} - \dfrac{20}{55}$

17) $\dfrac{41}{49} - \dfrac{36}{49}$

18) $\dfrac{40}{53} - \dfrac{39}{53}$

19) $\dfrac{27}{37} - \dfrac{17}{37}$

20) $\dfrac{21}{47} - \dfrac{11}{47}$

21) $\dfrac{24}{43} - \dfrac{12}{43}$

22) $\dfrac{13}{19} - \dfrac{12}{19}$

23) $\dfrac{6}{26} - \dfrac{3}{26}$

24) $\dfrac{9}{15} - \dfrac{7}{15}$

25) $\dfrac{8}{39} - \dfrac{3}{39}$

26) $\dfrac{18}{61} - \dfrac{15}{61}$

27) $\dfrac{12}{53} - \dfrac{9}{53}$

28) $\dfrac{75}{76} - \dfrac{74}{76}$

29) $\dfrac{26}{45} - \dfrac{13}{45}$

30) $\dfrac{20}{57} - \dfrac{17}{57}$

WWW.MathNotion.Com

GMAS Subject Test Mathematics Grade 5

Compare Fractions with Like Denominators

✏ Evaluate and compare. Write < or > or =.

1) $\frac{1}{3} + \frac{1}{3}$ ___ $\frac{1}{3}$

2) $\frac{3}{6} + \frac{3}{6}$ ___ $\frac{5}{6}$

3) $\frac{8}{9} - \frac{4}{9}$ ___ $\frac{7}{9}$

4) $\frac{4}{11} + \frac{5}{11}$ ___ $\frac{7}{11}$

5) $\frac{9}{14} - \frac{8}{14}$ ___ $\frac{5}{14}$

6) $\frac{11}{17} - \frac{3}{17}$ ___ $\frac{6}{17}$

7) $\frac{11}{21} + \frac{2}{21}$ ___ $\frac{10}{21}$

8) $\frac{8}{32} + \frac{6}{32}$ ___ $\frac{9}{32}$

9) $\frac{25}{29} - \frac{16}{29}$ ___ $\frac{11}{29}$

10) $\frac{28}{41} + \frac{13}{41}$ ___ $\frac{27}{41}$

11) $\frac{18}{35} - \frac{11}{35}$ ___ $\frac{22}{35}$

12) $\frac{32}{47} - \frac{22}{47}$ ___ $\frac{11}{47}$

13) $\frac{14}{27} + \frac{13}{27}$ ___ $\frac{24}{27}$

14) $\frac{34}{52} - \frac{11}{52}$ ___ $\frac{21}{52}$

15) $\frac{43}{56} - \frac{24}{56}$ ___ $\frac{27}{56}$

16) $\frac{27}{71} + \frac{25}{71}$ ___ $\frac{48}{71}$

GMAS Subject Test Mathematics Grade 5

More Than Two Fractions with Like Denominators

✎ Add fractions.

1) $\dfrac{5}{9} + \dfrac{2}{9} + \dfrac{2}{9}$

2) $\dfrac{4}{6} + \dfrac{1}{6} + \dfrac{1}{6}$

3) $\dfrac{2}{17} + \dfrac{4}{17} + \dfrac{2}{17}$

4) $\dfrac{1}{5} + \dfrac{1}{5} + \dfrac{1}{5}$

5) $\dfrac{7}{18} + \dfrac{2}{18} + \dfrac{3}{18}$

6) $\dfrac{3}{27} + \dfrac{5}{27} + \dfrac{2}{27}$

7) $\dfrac{4}{33} + \dfrac{4}{33} + \dfrac{4}{33}$

8) $\dfrac{8}{23} + \dfrac{6}{23} + \dfrac{2}{23}$

9) $\dfrac{13}{41} + \dfrac{2}{41} + \dfrac{8}{41}$

10) $\dfrac{6}{35} + \dfrac{9}{35} + \dfrac{20}{35}$

11) $\dfrac{1}{37} + \dfrac{5}{37} + \dfrac{5}{37}$

12) $\dfrac{4}{43} + \dfrac{9}{43} + \dfrac{8}{43}$

13) $\dfrac{4}{51} + \dfrac{10}{51} + \dfrac{7}{51}$

14) $\dfrac{5}{26} + \dfrac{13}{26} + \dfrac{6}{26}$

15) $\dfrac{5}{64} + \dfrac{4}{64} + \dfrac{2}{64}$

16) $\dfrac{1}{73} + \dfrac{5}{73} + \dfrac{6}{73}$

WWW.MathNotion.Com

GMAS Subject Test Mathematics Grade 5

Unlike Denominators

✏️ **Add fraction.**

1) $\dfrac{2}{9} + \dfrac{3}{4}$

2) $\dfrac{1}{4} + \dfrac{3}{5}$

3) $\dfrac{1}{16} + \dfrac{3}{4}$

4) $\dfrac{3}{8} + \dfrac{1}{7}$

5) $\dfrac{1}{3} + \dfrac{2}{4}$

6) $\dfrac{1}{6} + \dfrac{3}{7}$

7) $\dfrac{5}{18} + \dfrac{4}{6}$

8) $\dfrac{1}{12} + \dfrac{5}{6}$

9) $\dfrac{5}{27} + \dfrac{1}{9}$

10) $\dfrac{1}{6} + \dfrac{7}{24}$

11) $\dfrac{3}{5} + \dfrac{1}{8}$

12) $\dfrac{11}{42} + \dfrac{3}{7}$

13) $\dfrac{7}{20} + \dfrac{1}{3}$

14) $\dfrac{1}{45} + \dfrac{3}{5}$

15) $\dfrac{3}{32} + \dfrac{5}{8}$

16) $\dfrac{3}{48} + \dfrac{5}{6}$

17) $\dfrac{5}{12} + \dfrac{1}{6}$

18) $\dfrac{1}{34} + \dfrac{3}{17}$

19) $\dfrac{4}{9} + \dfrac{7}{54}$

20) $\dfrac{13}{56} + \dfrac{4}{7}$

21) $\dfrac{3}{12} + \dfrac{2}{3}$

22) $\dfrac{4}{33} + \dfrac{5}{11}$

GMAS Subject Test Mathematics Grade 5

✎ Subtract fractions.

1) $\dfrac{8}{9} - \dfrac{1}{2}$

2) $\dfrac{2}{3} - \dfrac{3}{10}$

3) $\dfrac{1}{6} - \dfrac{1}{9}$

4) $\dfrac{7}{8} - \dfrac{1}{4}$

5) $\dfrac{3}{4} - \dfrac{1}{28}$

6) $\dfrac{11}{30} - \dfrac{3}{15}$

7) $\dfrac{11}{18} - \dfrac{5}{9}$

8) $\dfrac{5}{13} - \dfrac{3}{26}$

9) $\dfrac{17}{35} - \dfrac{2}{7}$

10) $\dfrac{5}{6} - \dfrac{12}{36}$

11) $\dfrac{5}{9} - \dfrac{1}{27}$

12) $\dfrac{3}{5} - \dfrac{1}{8}$

13) $\dfrac{2}{3} - \dfrac{3}{5}$

14) $\dfrac{7}{8} - \dfrac{3}{7}$

15) $\dfrac{5}{9} - \dfrac{13}{45}$

16) $\dfrac{3}{4} - \dfrac{5}{36}$

17) $\dfrac{39}{49} - \dfrac{5}{7}$

18) $\dfrac{3}{11} - \dfrac{3}{22}$

19) $\dfrac{17}{48} - \dfrac{4}{12}$

20) $\dfrac{2}{3} - \dfrac{4}{13}$

21) $\dfrac{5}{8} - \dfrac{19}{72}$

22) $\dfrac{3}{5} - \dfrac{1}{12}$

GMAS Subject Test Mathematics Grade 5

Ordering Fractions

✎ Order the fractions from least to greatest.

1) $\frac{1}{5}, \frac{1}{11}, \frac{1}{8}, \frac{1}{3}$ ____, ____, ____, ____

2) $\frac{1}{9}, \frac{1}{18}, \frac{2}{4}, \frac{1}{5}$ ____, ____, ____, ____

3) $\frac{4}{7}, \frac{1}{7}, \frac{6}{21}, \frac{15}{21}$ ____, ____, ____, ____

4) $\frac{1}{2}, \frac{1}{3}, \frac{4}{9}, \frac{5}{18}$ ____, ____, ____, ____

5) $\frac{4}{9}, \frac{3}{4}, \frac{7}{36}, \frac{1}{6}$ ____, ____, ____, ____

✎ Order the fractions from greatest to least.

6) $\frac{3}{4}, \frac{4}{7}, \frac{3}{10}, \frac{5}{13}$ ____, ____, ____, ____

7) $\frac{5}{11}, \frac{5}{6}, \frac{2}{5}, \frac{1}{3}$ ____, ____, ____, ____

8) $\frac{7}{8}, \frac{1}{6}, \frac{3}{4}, \frac{5}{15}$ ____, ____, ____, ____

9) $\frac{4}{7}, \frac{2}{3}, \frac{11}{25}, \frac{13}{33}$ ____, ____, ____, ____

10) $\frac{18}{20}, \frac{15}{16}, \frac{14}{18}, \frac{5}{12}$ ____, ____, ____, ____

GMAS Subject Test Mathematics Grade 5

Denominators of 10, 100, and 1000

✎ Add fractions.

1) $\dfrac{7}{10} + \dfrac{13}{100}$

2) $\dfrac{1}{10} + \dfrac{10}{100}$

3) $\dfrac{15}{100} + \dfrac{1}{1,000}$

4) $\dfrac{56}{100} + \dfrac{3}{10}$

5) $\dfrac{50}{1,000} + \dfrac{7}{10}$

6) $\dfrac{6}{10} + \dfrac{30}{1,000}$

7) $\dfrac{9}{100} + \dfrac{3}{10}$

8) $\dfrac{5}{10} + \dfrac{50}{100}$

9) $\dfrac{48}{100} + \dfrac{6}{10}$

10) $\dfrac{70}{100} + \dfrac{2}{10}$

11) $\dfrac{80}{100} + \dfrac{200}{1,000}$

12) $\dfrac{30}{100} + \dfrac{4}{10}$

13) $\dfrac{9}{100} + \dfrac{7}{10}$

14) $\dfrac{25}{100} + \dfrac{6}{10}$

15) $\dfrac{15}{100} + \dfrac{8}{10}$

16) $\dfrac{3}{10} + \dfrac{31}{100}$

17) $\dfrac{8}{10} + \dfrac{11}{100}$

18) $\dfrac{34}{100} + \dfrac{6}{10}$

GMAS Subject Test Mathematics Grade 5

✏️ Subtract fractions.

1) $\dfrac{8}{10} - \dfrac{20}{100}$

2) $\dfrac{5}{10} - \dfrac{47}{100}$

3) $\dfrac{12}{100} - \dfrac{60}{1,000}$

4) $\dfrac{6}{10} - \dfrac{50}{100}$

5) $\dfrac{3}{10} - \dfrac{23}{100}$

6) $\dfrac{70}{100} - \dfrac{250}{1,000}$

7) $\dfrac{4}{10} - \dfrac{350}{1,000}$

8) $\dfrac{70}{100} - \dfrac{3}{10}$

9) $\dfrac{40}{100} - \dfrac{3}{10}$

10) $\dfrac{6}{10} - \dfrac{180}{1,000}$

11) $\dfrac{93}{100} - \dfrac{5}{10}$

12) $\dfrac{65}{100} - \dfrac{4}{10}$

13) $\dfrac{80}{100} - \dfrac{6}{10}$

14) $\dfrac{90}{100} - \dfrac{5}{10}$

15) $\dfrac{200}{1,000} - \dfrac{1}{10}$

16) $\dfrac{90}{100} - \dfrac{7}{10}$

17) $\dfrac{900}{1,000} - \dfrac{40}{100}$

18) $\dfrac{60}{100} - \dfrac{3}{10}$

Fractions to Mixed Numbers

✍ Convert fractions to mixed numbers.

1) $\dfrac{9}{5}$

2) $\dfrac{11}{3}$

3) $\dfrac{39}{8}$

4) $\dfrac{27}{11}$

5) $\dfrac{7}{2}$

6) $\dfrac{43}{4}$

7) $\dfrac{49}{9}$

8) $\dfrac{15}{4}$

9) $\dfrac{37}{7}$

10) $\dfrac{19}{7}$

11) $\dfrac{41}{9}$

12) $\dfrac{45}{12}$

13) $\dfrac{17}{5}$

14) $\dfrac{29}{6}$

15) $\dfrac{13}{4}$

16) $\dfrac{15}{7}$

17) $\dfrac{65}{7}$

18) $\dfrac{59}{8}$

19) $\dfrac{25}{4}$

20) $\dfrac{17}{8}$

Mixed Numbers to Fractions

✎ Convert to fraction.

1) $3\frac{3}{5}$

2) $1\frac{1}{3}$

3) $4\frac{2}{5}$

4) $4\frac{2}{8}$

5) $2\frac{1}{5}$

6) $2\frac{8}{11}$

7) $4\frac{4}{7}$

8) $3\frac{7}{12}$

9) $2\frac{1}{3}$

10) $7\frac{5}{7}$

11) $2\frac{7}{10}$

12) $3\frac{4}{9}$

13) $1\frac{5}{8}$

14) $4\frac{3}{11}$

15) $3\frac{4}{7}$

16) $5\frac{2}{8}$

17) $7\frac{1}{7}$

18) $13\frac{1}{2}$

19) $4\frac{2}{7}$

20) $5\frac{2}{10}$

21) $12\frac{1}{3}$

22) $7\frac{1}{8}$

GMAS Subject Test Mathematics Grade 5

Add and Subtract Mixed Numbers

✎ Add mixed numbers.

1) $3\frac{2}{5} + 8\frac{1}{5}$

2) $3\frac{2}{3} + 4\frac{1}{2}$

3) $6\frac{2}{7} + 2\frac{3}{7}$

4) $4\frac{2}{5} + 3\frac{1}{4}$

5) $8\frac{3}{4} - 2\frac{1}{2}$

6) $6\frac{5}{12} - 4\frac{1}{4}$

7) $5\frac{3}{8} - 3\frac{7}{8}$

8) $6\frac{1}{4} - 2\frac{15}{16}$

9) $9\frac{23}{28} - 4\frac{17}{28}$

10) $6\frac{1}{6} + 6\frac{2}{3}$

11) $4\frac{2}{9} + 5\frac{5}{9}$

12) $2\frac{1}{4} + 7\frac{4}{7}$

13) $7\frac{1}{5} - 3\frac{3}{5}$

14) $3\frac{1}{6} + 2\frac{3}{7}$

15) $2\frac{1}{3} + 4\frac{1}{4}$

16) $4\frac{1}{4} - 1\frac{2}{5}$

17) $\frac{1}{3} + 6\frac{1}{6}$

18) $2\frac{3}{5} + 2\frac{1}{10}$

WWW.MathNotion.Com

Answers of Worksheets

Simplifying Fractions

1) $\frac{11}{21}$
2) $\frac{2}{5}$
3) $\frac{3}{4}$
4) $\frac{1}{6}$
5) $\frac{1}{2}$
6) $\frac{1}{7}$
7) $\frac{2}{7}$
8) $\frac{1}{3}$
9) $\frac{4}{5}$
10) $\frac{1}{7}$
11) $\frac{1}{3}$
12) $\frac{3}{2}$
13) $\frac{5}{7}$
14) $\frac{3}{4}$
15) $\frac{1}{5}$
16) $\frac{5}{8}$
17) $\frac{3}{8}$
18) $\frac{3}{8}$
19) $\frac{1}{9}$
20) $\frac{3}{4}$
21) $\frac{1}{9}$
22) $\frac{1}{2}$

Like Denominators (addition)

1) 1
2) 1
3) $\frac{11}{9}$
4) $\frac{4}{7}$
5) $\frac{7}{13}$
6) $\frac{5}{14}$
7) $\frac{12}{19}$
8) $\frac{3}{4}$
9) $\frac{2}{5}$
10) $\frac{8}{17}$
11) $\frac{5}{11}$
12) $\frac{18}{35}$
13) 1
14) $\frac{12}{31}$
15) $\frac{8}{23}$
16) $\frac{21}{41}$
17) $\frac{33}{37}$
18) $\frac{3}{17}$
19) $\frac{23}{26}$
20) $\frac{23}{48}$
21) $\frac{19}{29}$
22) 1
23) $\frac{6}{19}$
24) $\frac{7}{53}$
25) $\frac{9}{20}$
26) $\frac{8}{63}$
27) $\frac{7}{38}$
28) 1
29) $\frac{2}{7}$
30) $\frac{17}{37}$

Like Denominators (Subtraction)

1) $\frac{4}{9}$
2) $\frac{1}{4}$
3) $\frac{6}{11}$
4) $\frac{5}{14}$
5) $\frac{7}{20}$
6) $\frac{1}{15}$
7) $\frac{2}{19}$
8) $\frac{3}{4}$
9) $\frac{3}{29}$
10) $\frac{7}{23}$
11) $\frac{8}{34}$
12) $\frac{9}{41}$
13) $\frac{1}{39}$
14) $\frac{2}{13}$
15) $\frac{10}{17}$
16) $\frac{13}{55}$
17) $\frac{5}{49}$
18) $\frac{1}{53}$
19) $\frac{10}{37}$
20) $\frac{10}{47}$
21) $\frac{12}{43}$
22) $\frac{1}{19}$
23) $\frac{3}{26}$
24) $\frac{2}{15}$

GMAS Subject Test Mathematics Grade 5

25) $\frac{5}{39}$ 27) $\frac{3}{53}$ 29) $\frac{13}{45}$

26) $\frac{3}{61}$ 28) $\frac{1}{76}$ 30) $\frac{1}{19}$

Compare Fractions with Like Denominators

1) $\frac{2}{3} > \frac{1}{3}$ 5) $\frac{1}{14} < \frac{5}{14}$ 9) $\frac{9}{29} < \frac{11}{29}$ 13) $1 > \frac{24}{27}$

2) $1 > \frac{5}{6}$ 6) $\frac{8}{17} > \frac{6}{17}$ 10) $1 > \frac{27}{41}$ 14) $\frac{23}{52} > \frac{21}{52}$

3) $\frac{4}{9} < \frac{7}{9}$ 7) $\frac{13}{21} > \frac{10}{21}$ 11) $\frac{7}{35} < \frac{22}{35}$ 15) $\frac{19}{56} < \frac{27}{56}$

4) $\frac{9}{11} > \frac{7}{11}$ 8) $\frac{14}{32} > \frac{9}{32}$ 12) $\frac{10}{47} < \frac{11}{47}$ 16) $\frac{52}{71} > \frac{48}{71}$

More Than Two Fractions with Like Denominators

1) 1   5) $\frac{2}{3}$ 9) $\frac{23}{41}$ 13) $\frac{7}{17}$

2) 1   6) $\frac{10}{27}$ 10) 1   14) $\frac{12}{13}$

3) $\frac{8}{17}$ 7) $\frac{4}{11}$ 11) $\frac{11}{37}$ 15) $\frac{11}{64}$

4) $\frac{3}{5}$ 8) $\frac{16}{23}$ 12) $\frac{21}{43}$ 16) $\frac{12}{73}$

Unlike Denominators (Addition)

1) $\frac{35}{36}$ 7) $\frac{17}{18}$ 13) $\frac{41}{60}$ 19) $\frac{31}{54}$

2) $\frac{17}{20}$ 8) $\frac{11}{12}$ 14) $\frac{28}{45}$ 20) $\frac{45}{56}$

3) $\frac{13}{16}$ 9) $\frac{8}{27}$ 15) $\frac{23}{32}$ 21) $\frac{11}{12}$

4) $\frac{29}{56}$ 10) $\frac{11}{24}$ 16) $\frac{43}{48}$ 22) $\frac{19}{33}$

5) $\frac{5}{6}$ 11) $\frac{29}{40}$ 17) $\frac{7}{12}$

6) $\frac{25}{42}$ 12) $\frac{29}{42}$ 18) $\frac{7}{34}$

Unlike Denominators (Subtraction)

1) $\frac{7}{18}$ 5) $\frac{5}{7}$ 9) $\frac{1}{5}$ 13) $\frac{1}{15}$

2) $\frac{11}{30}$ 6) $\frac{1}{6}$ 10) $\frac{1}{2}$ 14) $\frac{25}{56}$

3) $\frac{1}{18}$ 7) $\frac{1}{18}$ 11) $\frac{14}{27}$ 15) $\frac{4}{15}$

4) $\frac{5}{8}$ 8) $\frac{7}{26}$ 12) $\frac{19}{40}$ 16) $\frac{11}{18}$

WWW.MathNotion.Com

GMAS Subject Test Mathematics Grade 5

17) $\frac{4}{49}$ 19) $\frac{1}{48}$ 21) $\frac{13}{36}$

18) $\frac{3}{22}$ 20) $\frac{14}{39}$ 22) $\frac{31}{60}$

Ordering Fractions

1) $\frac{1}{11}, \frac{1}{8}, \frac{1}{5}, \frac{1}{3}$ 5) $\frac{1}{6}, \frac{7}{36}, \frac{4}{9}, \frac{3}{4}$ 9) $\frac{2}{3}, \frac{4}{7}, \frac{11}{25}, \frac{13}{33}$

2) $\frac{1}{18}, \frac{1}{9}, \frac{1}{5}, \frac{2}{4}$ 6) $\frac{3}{4}, \frac{4}{7}, \frac{5}{13}, \frac{3}{10}$ 10) $\frac{15}{16}, \frac{18}{20}, \frac{14}{18}, \frac{5}{12}$

3) $\frac{1}{7}, \frac{6}{21}, \frac{4}{7}, \frac{15}{21}$ 7) $\frac{5}{6}, \frac{5}{11}, \frac{2}{5}, \frac{1}{3}$

4) $\frac{5}{18}, \frac{1}{3}, \frac{4}{9}, \frac{1}{2}$ 8) $\frac{7}{8}, \frac{3}{4}, \frac{5}{15}, \frac{1}{6}$

Denominators of 10, 100, and 1000

1) $\frac{83}{100}$ 6) $\frac{63}{100}$ 11) 1   16) $\frac{61}{100}$

2) $\frac{1}{5}$ 7) $\frac{39}{100}$ 12) $\frac{7}{10}$ 17) $\frac{91}{100}$

3) $\frac{151}{1,000}$ 8) 1   13) $\frac{79}{100}$ 18) $\frac{47}{50}$

4) $\frac{43}{50}$ 9) $\frac{27}{25}$ 14) $\frac{17}{20}$

5) $\frac{3}{4}$ 10) $\frac{9}{10}$ 15) $\frac{19}{20}$

Denominators of 10, 100, and 1000 (Subtract)

1) $\frac{3}{5}$ 6) $\frac{9}{20}$ 11) $\frac{43}{100}$ 16) $\frac{1}{5}$

2) $\frac{3}{100}$ 7) $\frac{1}{20}$ 12) $\frac{1}{4}$ 17) $\frac{1}{2}$

3) $\frac{3}{50}$ 8) $\frac{2}{5}$ 13) $\frac{1}{5}$ 18) $\frac{3}{10}$

4) $\frac{1}{10}$ 9) $\frac{1}{10}$ 14) $\frac{2}{5}$

5) $\frac{7}{100}$ 10) $\frac{21}{50}$ 15) $\frac{1}{10}$

Fractions to Mixed Numbers

1) $1\frac{4}{5}$ 5) $3\frac{1}{2}$ 9) $5\frac{2}{7}$ 13) $3\frac{2}{5}$

2) $3\frac{2}{3}$ 6) $10\frac{3}{4}$ 10) $2\frac{5}{7}$ 14) $4\frac{5}{6}$

3) $4\frac{7}{8}$ 7) $5\frac{4}{9}$ 11) $4\frac{5}{9}$ 15) $3\frac{1}{4}$

4) $2\frac{5}{11}$ 8) $3\frac{3}{4}$ 12) $3\frac{9}{12}$ 16) $2\frac{1}{7}$

17) $9\frac{2}{7}$ 18) $7\frac{3}{8}$ 19) $6\frac{1}{4}$ 20) $2\frac{1}{8}$

Mixed Numbers to Fractions

1) $\frac{18}{5}$ 7) $\frac{32}{7}$ 13) $\frac{13}{8}$ 19) $\frac{30}{7}$

2) $\frac{4}{3}$ 8) $\frac{43}{12}$ 14) $\frac{47}{11}$ 20) $\frac{52}{10}$

3) $\frac{22}{5}$ 9) $\frac{7}{3}$ 15) $\frac{25}{7}$ 21) $\frac{37}{3}$

4) $\frac{34}{8}$ 10) $\frac{54}{7}$ 16) $\frac{42}{8}$ 22) $\frac{57}{8}$

5) $\frac{11}{5}$ 11) $\frac{27}{10}$ 17) $\frac{50}{7}$

6) $\frac{30}{11}$ 12) $\frac{31}{9}$ 18) $\frac{27}{2}$

Add and Subtract Mixed Numbers

1) $11\frac{3}{5}$ 6) $2\frac{1}{6}$ 11) $9\frac{7}{9}$ 16) $2\frac{17}{20}$

2) $8\frac{1}{6}$ 7) $1\frac{1}{2}$ 12) $9\frac{23}{28}$ 17) $6\frac{1}{2}$

3) $8\frac{5}{7}$ 8) $3\frac{5}{16}$ 13) $3\frac{3}{5}$ 18) $4\frac{7}{10}$

4) $7\frac{13}{20}$ 9) $5\frac{3}{14}$ 14) $5\frac{25}{42}$

5) $6\frac{1}{4}$ 10) $12\frac{5}{6}$ 15) $6\frac{7}{12}$

Chapter 5 : Decimals

Topics that you'll learn in this chapter:

- ✓ Adding and Subtracting Decimals,
- ✓ Multiplying and Dividing Decimals,
- ✓ Round decimals,
- ✓ Comparing Decimals,

GMAS Subject Test Mathematics Grade 5

Adding and Subtracting Decimals

Add and subtract decimals.

1) 24.19 − 15.42

2) 42.23 + 25.42

3) 72.54 + 11.28

4) 57.45 − 24.75

5) 43.57 + 54.85

6) 86.68 − 54.12

Solve.

7) ____ + 2.7 = 8.1

8) 6.4 + ____ = 12.8

9) 7.9 + ____ = 17

10) 4.6 + ____ = 15.3

11) ____ + 9.4 = 15

12) ____ + 8.24 = 13.54

Order each set of numbers from least to greatest.

1) 0.4, 0.67, 0.44, 0.73, 0.51   ___, ___, ___, ___, ___, ___

2) 3.9, 6.1, 4.28, 7.02, 4.65   ___, ___, ___, ___, ___, ___

3) 1.9, 1.04, 0.79, 0.72, 0.09   ___, ___, ___, ___, ___, ___

4) 2.6, 5.2, 1.9, 4.01, 1.99, 3.2   ___, ___, ___, ___, ___, ___

5) 4.2, 6.1, 3.8, 5.7, 2.1, 2.8   ___, ___, ___, ___, ___, ___

6) 0.56, 0.87, 0.14, 1.24, 3.1   ___, ___, ___, ___, ___, ___

GMAS Subject Test Mathematics Grade 5

Multiplying and Dividing Decimals

✎ Find each product.

1) 1.5
   × 2.1
   _____

2) 4.6
   × 3.4
   _____

3) 6.3
   × 2.5
   _____

4) 6.5
   × 0.99
   _____

5) 12.1
   × 5.2
   _____

6) 3.4
   × 8.9
   _____

7) 4.8
   × 9.1
   _____

8) 22.35
   × 20
   _____

9) 15.25
   × 3.6
   _____

✎ Find each quotient.

10) 3.5 ÷ 0.85

11) 15.35 ÷ 4.6

12) 32.42 ÷ 8.8

13) 9.2 ÷ 3.4

14) 0.84 ÷ 0.1

15) 21.5 ÷ 1,000

16) 4.1 ÷ 100

17) 9.7 ÷ 10

18) 6.55 ÷ 1.25

19) 18.48 ÷ 11.2

WWW.MathNotion.Com

GMAS Subject Test Mathematics Grade 5

Rounding Decimals

✎ Round each decimal number to the nearest place indicated.

1) 0.3<u>2</u>

2) 5.<u>0</u>1

3) 8.<u>8</u>24

4) 0.<u>4</u>78

5) <u>7</u>.32

6) 0.<u>2</u>9

7) 11.<u>3</u>1

8) <u>6</u>.223

9) 9.6<u>3</u>7

10) 5.<u>4</u>804

11) <u>7</u>.9

12) <u>5</u>.2439

13) 6.<u>4</u>92

14) 1.<u>6</u>2

15) 7<u>2</u>.85

16) 8<u>3</u>.67

17) 41.<u>6</u>8

18) 79<u>4</u>.741

19) 5<u>2</u>.2

20) 7<u>6</u>.93

21) <u>3</u>.219

22) 7<u>2</u>.09

23) 486.<u>4</u>91

24) 7.<u>0</u>8

Comparing Decimals

✎ Write the correct comparison symbol (>, < or =).

1) 0.35 __ 1.8

2) 1.9 __ 1.19

3) 8.6 __ 8.6

4) 2.45 __ 24.5

5) 7.56 __ 0.756

6) 11.4 __ 11.05

7) 7.4 __ 0.7.4

8) 8.56 __ 0.85

9) 7 __ 0.7

10) 7.12 __ 0.712

11) 12.3 __ 12.5

12) 4.67 __ 4.68

13) 2.57 __ 2.75

14) 3.46 __ 0.346

15) 6.87 __ 6.78

16) 0.89 __ 0.98

17) 1.57 __ 0.157

18) 0.092 __ 0.091

19) 24.3 __ 24.3

20) 0.17 __ 0.71

21) 0.46 __ 0.64

22) 0.2 __ 0.08

23) 0.10 __ 0.1

24) 3.52 __ 31.5

GMAS Subject Test Mathematics Grade 5

Answers of Worksheets

Adding and Subtracting Decimals

1) 8.77   4) 32.7   7) 5.4   10) 10.7
2) 67.65   5) 98.42   8) 6.4   11) 5.6
3) 83.82   6) 32.56   9) 9.1   12) 5.3

Order and Comparing Decimals

1) 0.4, 0.44, 0.51, 0.67, 0.73   4) 1.9, 1.99, 2.6, 3.2, 4.01, 5.2
2) 3.9, 4.28, 4.65, 6.1, 7.02   5) 2.1, 2.8, 3.8, 4.2, 5.7, 6.1
3) 0.09, 0.72, 0.79, 1.04, 1.9   6) 0.14, 0.56, 0.87, 1.24, 3.1

Multiplying and Dividing Decimals

1) 3.15   6) 30.26   11) 3.336…   16) 0.041
2) 15.64   7) 43.68   12) 3.684…   17) 0.97
3) 15.75   8) 447   13) 2.705…   18) 5.24
4) 6.435   9) 54.9   14) 8.4   19) 1.65
5) 62.92   10) 4.117…   15) 0.0215

Rounding Decimals

1) 0.3   7) 11.3   13) 6.5   19) 52
2) 5.0   8) 6   14) 1.6   20) 77
3) 8.8   9) 9.64   15) 73   21) 3
4) 0.5   10) 5.5   16) 84   22) 72
5) 7   11) 8   17) 41.7   23) 486.5
6) 0.3   12) 5   18) 795   24) 7.1

Comparing Decimals

1) 0.35 < 1.8   9) 7 > 0.7   17) 1.57 > 0.157
2) 1.9 > 1.19   10) 7.12 > 0.712   18) 0.092 > 0.091
3) 8.6 = 8.6   11) 12.3 < 12.5   19) 24.3 = 24.3
4) 2.45 < 24.5   12) 4.67 < 4.68   20) 0.17 < 0.71
5) 7.56 > 0.756   13) 2.57 < 2.75   21) 0.46 < 0.64
6) 11.4 > 11.05   14) 3.46 > 0.346   22) 0.2 > 0.08
7) 7.4 > 0.74   15) 6.87 > 6.78   23) 0.10 = 0.1
8) 8.56 > 0.85   16) 0.89 < 0.98   24) 3.52 < 31.5

WWW.MathNotion.Com

Chapter 6 : Ratios and Rates

Topics that you'll learn in this chapter:

- ✓ Simplifying Ratios,
- ✓ Writing Ratios,
- ✓ Create a Proportion,
- ✓ Proportional Ratios,
- ✓ Similar Figures,
- ✓ Word Problems,

GMAS Subject Test Mathematics Grade 5

Simplifying Ratios

✏️ Reduce each ratio.

1) 14: 56
2) 12: 36
3) 5: 35
4) 56: 48
5) 12: 14
6) 81: 63
7) 60: 3
8) 15: 10
9) 25: 20
10) 14: 28
11) 60: 84
12) 11: 99
13) 30: 45
14) 18: 45
15) 90: 15
16) 1.5: 3
17) 8: 88
18) 13: 52
19) 3: 75
20) 2.2: 22
21) 11: 33
22) 18: 81
23) 68: 80
24) 50: 500

Writing Ratios

✏️ Express each ratio as a rate and unite rate.

1) 180 miles on 6 gallons of gas.

2) 99 dollars for 11 books.

3) 35 miles on 3.5 gallons of gas

4) 7.5 inches of snow in 1.5 hours

✏️ Express each ratio as a fraction in the simplest form.

5) 6 feet out of 60 feet
6) 14 cakes out of 49 cakes
7) 32 dimes t0 60 dimes
8) 18 dimes out of 63 coins
9) 13 cups to 91 cups
10) 28 gallons to 42 gallons
11) 35 miles out of 120 miles
12) 22 blue cars out of 55 cars
13) 6.9 pennies to 69 pennies
14) 14 beetles out of 70 insects
15) 18 dimes to 54 dimes
16) 40 red cars out of 160 cars

GMAS Subject Test Mathematics Grade 5

Create a Proportion

✎ Create proportion from the given set of numbers.

1) 1, 20, 4, 5

2) 9, 135, 1, 15

3) 5, 15, 8, 24

4) 49, 7, 4, 28

5) 9, 1, 108, 12

6) 45, 2, 5, 18

7) 28, 7, 24, 6

8) 11, 3, 55, 15

9) 5, 45, 36, 4

10) 16, 128, 1, 8

11) 2.5, 10, 5, 20

12) 5, 12, 15, 36

Proportional Ratios

✎ Solve each proportion.

1) $\frac{4}{8} = \frac{5}{d}$

2) $\frac{k}{6} = \frac{3}{9}$

3) $\frac{10}{8} = \frac{12}{x}$

4) $\frac{x}{15} = \frac{9}{5}$

5) $\frac{d}{11} = \frac{10}{1.1}$

6) $\frac{4.5}{6} = \frac{9}{x}$

7) $\frac{7}{15} = \frac{k}{60}$

8) $\frac{11}{1.5} = \frac{121}{d}$

9) $\frac{x}{0.7} = \frac{15}{2.8}$

10) $\frac{1.2}{4} = \frac{x}{2.5}$

11) $\frac{7.8}{x} = \frac{7.8}{2}$

12) $\frac{x}{3.4} = \frac{48}{16}$

13) $\frac{80}{20} = \frac{k}{60}$

14) $\frac{1.4}{5} = \frac{28}{d}$

15) $\frac{x}{7} = \frac{30}{15}$

16) $\frac{4}{1.6} = \frac{k}{1.6}$

17) $\frac{0.8}{1.2} = \frac{5.6}{d}$

18) $\frac{25}{x} = \frac{50}{4}$

19) $\frac{d}{9} = \frac{18}{27}$

20) $\frac{k}{12.6} = \frac{5}{12.6}$

21) $\frac{1.4}{7} = \frac{x}{10}$

22) $\frac{13}{5} = \frac{k}{15}$

23) $\frac{18}{21} = \frac{x}{7}$

24) $\frac{9}{99} = \frac{x}{22}$

WWW.MathNotion.Com

GMAS Subject Test Mathematics Grade 5

Similar Figures

🖋 Each pair of figures is similar. Find the missing side.

1)

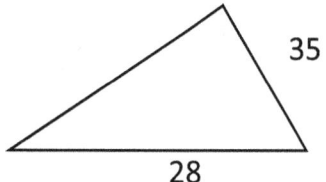 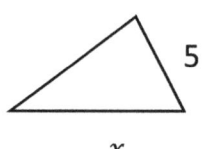

2)

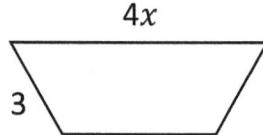 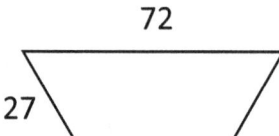

3)

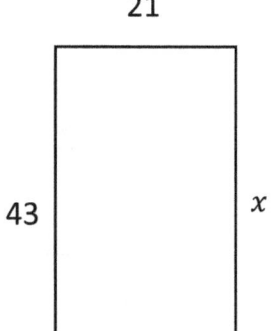 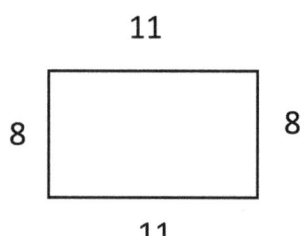

WWW.MathNotion.Com

Word Problems

✎ Solve.

1) In a party, 15 soft drinks are required for every 18 guests. If there are 360 guests, how many soft drinks is required?

2) In Jack's class, 16 of the students are tall and 10 are short. In Michael's class 40 students are tall and 25 students are short. Which class has a higher ratio of tall to short students?

3) Are these ratios equivalent?

12 cards to 84 animals 18 marbles to 126 marbles

4) The price of 6 apples at the Quick Market is $2.7. The price of 7 of the same apples at Walmart is $3.64. Which place is the better buy?

5) The bakers at a Bakery can make 160 bagels in 4 hours. How many bagels can they bake in 11 hours? What is that rate per hour?

GMAS Subject Test Mathematics Grade 5

✎ Answer each question and round your answer to the nearest whole number.

6) If a 24.6 ft tall flagpole casts a 190.95 ft long shadow, then how long is the shadow that a 4.7 ft tall woman casts?

7) A model igloo has a scale of 2 in: 7 ft. If the real igloo is 56 ft wide, then how wide is the model igloo?

8) If an 88 ft tall tree casts a 8 ft long shadow, then how tall is an adult giraffe that casts a 4 ft shadow?

9) Find the distance between San Joe and Mount Pleasant if they are 5 cm apart on a map with a scale of 1 cm: 8 km.

10) A telephone booth that is 54 ft tall casts a shadow that is 9 ft long. Find the height of a lawn ornament that casts a 7 ft shadow.

GMAS Subject Test Mathematics Grade 5

Answers of Worksheets

Simplifying Ratios

1) 2: 8	7) 20: 1	13) 2: 3	19) 1: 25
2) 1: 3	8) 3: 2	14) 2: 5	20) 1: 10
3) 1: 7	9) 5: 4	15) 6: 1	21) 1: 3
4) 7: 6	10) 1: 2	16) 1: 2	22) 2: 9
5) 6: 7	11) 5: 7	17) 1: 11	23) 17: 20
6) 9: 7	12) 1: 9	18) 1: 4	24) 1: 10

Writing Ratios

1) $\frac{180 \text{ miles}}{6 \text{ gallons}}$, 30 miles per gallon

2) $\frac{99 \text{ dollars}}{11 \text{ books}}$, 9.00 dollars per book

3) $\frac{35 \text{ miles}}{3.5 \text{ gallons}}$, 10 miles per gallon

4) $\frac{7.5" \text{ of snow}}{1.5 \text{ hours}}$, 5 inches of snow per hour

5) $\frac{1}{10}$	8) $\frac{2}{7}$	11) $\frac{7}{24}$	14) $\frac{1}{5}$
6) $\frac{2}{7}$	9) $\frac{1}{7}$	12) $\frac{2}{5}$	15) $\frac{1}{3}$
7) $\frac{8}{15}$	10) $\frac{2}{3}$	13) $\frac{1}{10}$	16) $\frac{1}{4}$

Create Proportion

1) 1: 5 = 4: 20	5) 9: 1 = 108: 12	9) 4: 5 = 36: 45
2) 9: 135 = 1: 15	6) 45: 18 = 5: 2	10) 128: 16 = 8: 1
3) 8: 5 = 24: 15	7) 24: 28 = 6: 7	11) 2.5: 5 = 10: 20
4) 7: 4 = 49: 28	8) 11: 3 = 55: 15	12) 15: 5 = 36: 12

Proportional Ratios

1) 10	7) 28	13) 240	19) 6
2) 2	8) 16.5	14) 100	20) 5
3) 9.6	9) 3.75	15) 14	21) 2
4) 27	10) 0.75	16) 4	22) 39
5) 100	11) 2	17) 8.4	23) 6
6) 12	12) 10.2	18) 2	24) 2

Similar Figures

1) 4   2) 2   3) 43

GMAS Subject Test Mathematics Grade 5

Word Problems

1) 300

2) The ratio for both classes is equal to 8 to 5.

3) Yes! Both ratios are 1 to 7

4) The price at the Quick Market is a better buy.

5) 440, the rate is 40 per hour.

6) 36.48 ft   8) 44 ft   10) 42 ft

7) 16 in   9) 40 km

Chapter 7 : Measurement

Topics that you'll learn in this chapter:

- ✓ Reference Measurement Units,
- ✓ Metric Length Units,
- ✓ Customary Length Units,
- ✓ Metric Capacity Units,
- ✓ Customary Capacity Units,
- ✓ Metric Weight and Mass Units,
- ✓ Customary Weight and Mass Units,
- ✓ Temperature Units,
- ✓ Time,
- ✓ Add Money Amounts,
- ✓ Subtract Money Amounts,
- ✓ Money: Word Problems,

GMAS Subject Test Mathematics Grade 5

Reference Measurement Units

LENGTH

Customary

1 mile (mi) = 1,760 yards (yd)

1 yard (yd) = 3 feet (ft)

1 foot (ft) = 12 inches (in.)

Metric

1 kilometer (km) = 1,000 meters (m)

1 meter (m) = 100 centimeters (cm)

1 centimeter(cm)= 10 millimeters(mm)

VOLUME AND CAPACITY

Customary

1 gallon (gal) = 4 quarts (qt)

1 quart (qt) = 2 pints (pt.)

1 pint (pt.) = 2 cups (c)

1 cup (c) = 8 fluid ounces (Fl oz)

Metric

1 liter (L) = 1,000 milliliters (mL)

WEIGHT AND MASS

Customary

1 ton (T) = 2,000 pounds (lb.)

1 pound (lb.) = 16 ounces (oz)

Metric

1 kilogram (kg) = 1,000 grams (g)

1 gram (g) = 1,000 milligrams (mg)

Time

1 year = 12 months

1 year = 52 weeks

1 week = 7 days

1 day = 24 hours

1 hour = 60 minutes

1 minute = 60 seconds

GMAS Subject Test Mathematics Grade 5

Metric Length Units

✏️ Convert to the units.

1) 300 mm = _____ cm

2) 8 m = _____ mm

3) 4.5 m = _____ cm

4) 7 km = _____ m

5) 9,400 mm = _____ m

6) 1,100 cm = _____ m

7) 2.8 m = _____ cm

8) 4,000 mm = _____ cm

9) 7,000 mm = _____ m

10) 2 km = _____ mm

11) 14.9 km = _____ m

12) 20 m = _____ cm

13) 5,000 m = _____ km

14) 7,600 m = _____ km

Customary Length Units

✏️ Convert to the units.

1) 8 ft = _____ in

2) 4 ft = _____ in

3) 6 yd = _____ ft

4) 10 yd = _____ ft

5) 3,520 yd = _____ mi

6) 60 in = _____ ft

7) 144 in = _____ yd

8) 0.5 mi = _____ yd

9) 15 yd = _____ in

10) 42 yd = _____ in

11) 99 ft = _____ yd

12) 1.5 mi = _____ yd

13) 84 in = _____ ft

14) 30 yd = _____ feet

GMAS Subject Test Mathematics Grade 5

Metric Capacity Units

✎Convert the following measurements.

1) 32.4 l = _____ ml

2) 7.1 l = _____ ml

3) 54 l = _____ ml

4) 92 l = _____ ml

5) 48 l = _____ ml

6) 13 l = _____ ml

7) 750 ml = _____ l

8) 2,400 ml = _____ l

9) 73,000 ml = _____ l

10) 8,000 ml = _____ l

11) 49,000 ml = _____ l

12) 5,500 ml = _____ l

Customary Capacity Units

✎Convert the following measurements.

1) 51 gal = _____ qt.

2) 35 gal = _____ pt.

3) 68 gal = _____ c.

4) 20 pt. = _____ c

5) 12.5 qt = _____ pt.

6) 22.5 qt = _____ c

7) 51 pt. = _____ c

8) 48 c = _____ gal

9) 96 pt. = _____ gal

10) 136 qt = _____ gal

11) 15 c = _____ fl oz

12) 44 c = _____ qt

13) 240 c = _____ pt.

14) 148 qt = _____ gal

15) 160 pt. = _____ qt

16) 104 fl oz = _____ c.

WWW.MathNotion.Com

GMAS Subject Test Mathematics Grade 5

Metric Weight and Mass Units

☑ Convert.

1) 60 kg = _____ g

2) 24 kg = _____ g

3) 610 kg = _____ g

4) 82 kg = _____ g

5) 95.8 kg = _____ g

6) 4.85 kg = _____ g

7) 1.5 kg = _____ g

8) 95,000 g = _____ kg

9) 241,000 g = _____ kg

10) 700,000 g = _____ kg

11) 2,500 g = _____ kg

12) 28,900 g = _____ kg

13) 970,000 g = _____ kg

14) 325,500 g = _____ kg

Customary Weight and Mass Units

☑ Convert.

1) 10,000 lb. = _____ T

2) 19,000 lb. = _____ T

3) 32,000 lb. = _____ T

4) 16,800 lb. = _____ T

5) 27 lb. = _____ oz

6) 25.4 lb. = _____ oz

7) 124 lb. = _____ oz

8) 4T = _____ lb.

9) 7T = _____ lb.

10) 11.2T = _____ lb.

11) 12.8T = _____ lb.

12) $\frac{6}{5}$ T = _____ oz

13) 9.125 T = _____ oz

14) $\frac{3}{4}$ T = _____ oz

www.MathNotion.Com

Temperature Units

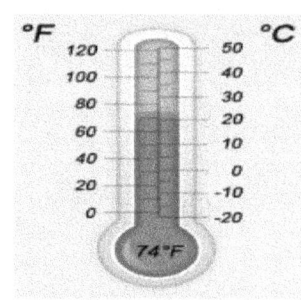

✎ Convert Fahrenheit into Celsius.

1) 21°F = ___ °C

2) 5.5°F = ___ °C

3) 71.6°F = ___ °C

4) 248°F = ___ °C

5) 75.2°F = ___ °C

6) 138.2°F = ___ °C

7) 167°F = ___ °C

8) 215.6°F = ___ °C

9) 104°F = ___ °C

10) 194°F = ___ °C

11) 203°F = ___ °C

12) 78.8°F = ___ °C

✎ Convert Celsius into Fahrenheit.

13) 2°C = ___ °F

14) 19°C = ___ °F

15) 48°C = ___ °F

16) 54°C = ___ °F

17) 81°C = ___ °F

18) 24°C = ___ °F

19) 88°C = ___ °F

20) 72°C = ___ °F

21) 80°C = ___ °F

22) 16°C = ___ °F

23) 10°C = ___ °F

24) 110°C = ___ °F

GMAS Subject Test Mathematics Grade 5

Time

✏ Convert to the units.

1) 22 hr. = _____ min

2) 12.5 year = _____ week

3) 5.2 hr = _____ sec

4) 21 min = _____ sec

5) 1,800 min = _____ hr.

6) 730 day = _____ year

7) 1.5 year = _____ hr.

8) 42 day = _____ hr.

9) 2 day = _____ min

10) 660 min = _____ hr.

11) 10 year = _____ month

12) 2,124 sec = _____ min

13) 168 hr = _____ day

14) 19 weeks = _____ day

✏ How much time has passed?

1) From 2:25 A.M. to 5:35 A.M.: ____ hours and ____ minutes.

2) From 3:30 A.M. to 7:55 A.M.: ____ hours and ____ minutes.

3) It's 6:30 P.M. What time was 2 hours ago? _____ O'clock

4) 4:30 A.M to 7:50 AM: ____ hours and ____ minutes.

5) 4:15 A.M to 8:35 AM: ____ hours and ____ minutes.

6) 5:10 A.M. to 7:35 AM. = ____ hour(s) and ____ minutes.

7) 10:55 A.M. to 3:25 PM. = ____ hour(s) and ____ minutes

8) 8:05 A.M. to 8:40 A.M. = _____ minutes

9) 6:02 A.M. to 6:49 A.M. = _____ minutes

WWW.MathNotion.Com

GMAS Subject Test Mathematics Grade 5

Money Amounts

✎ Add.

1) $128 $225 $150
   +$328   +$245   +$186
   _____   _____   _____

2) $453 $440 $258
   +$128   +$541   +$248
   _____   _____   _____

3) $645 $235.4 $125.99
   +$112.5   +$452.1   +$148.32
   _____   _____   _____

4) $321.40 $458.10 $652.00
   +$175.80   +$752.65   +$324.70
   _____   _____   _____

✎ Subtract.

5) $725 $543 $349
   −$334   −$248   −$122
   _____   _____   _____

6) $658.20 $752.10 $312.50
   −$220.30   −$452.15   −$89.90
   _____   _____   _____

7) $315.90 $548.40 $968.40
   −$220.10   −$342.10   −$324.50
   _____   _____   _____

8) Linda had $18.60. She bought some game tickets for $9.25. How much did she have left?

Money: Word Problems

🖉 Solve.

1) How many boxes of envelopes can you buy with $40 if one box costs $8?

2) After paying $7.22 for a salad, Ella has $45.86. How much money did she have before buying the salad?

3) How many packages of diapers can you buy with $96 if one package costs $6?

4) Last week James ran 28.5 miles more than Michael. James ran 59 miles. How many miles did Michael run?

5) Last Friday Jacob had $14.68. Over the weekend he received some money for cleaning the attic. He now has $38.95. How much money did he receive?

6) After paying $3.15 for a sandwich, Amelia has $48.69. How much money did she have before buying the sandwich?

GMAS Subject Test Mathematics Grade 5

Answers of Worksheets

Metric length

1) 30 cm
2) 8,000 mm
3) 450 cm
4) 7,000 m
5) 9.4 m
6) 11 m
7) 280 cm
8) 400 cm
9) 7 m
10) 2,000,000 mm
11) 14,900 m
12) 2,000 cm
13) 5 km
14) 7.6 km

Customary Length

1) 96
2) 48
3) 18
4) 30
5) 2
6) 5
7) 4
8) 880
9) 540
10) 1,512
11) 33
12) 2,640
13) 7
14) 90

Metric Capacity

1) 32,400 ml
2) 7,100 ml
3) 54,000 ml
4) 92,000 ml
5) 48,000 ml
6) 13,000 ml
7) 0.75ml
8) 2.4 ml
9) 73 ml
10) 8L
11) 49 L
12) 5.5 L

Customary Capacity

1) 204 qt
2) 280 pt.
3) 1,088 c
4) 40 c
5) 25 pt.
6) 90c
7) 102 c
8) 3 gal
9) 12 gal
10) 34 gal
11) 120 qt
12) 11qt
13) 120 pt.
14) 37 gal
15) 80 qt
16) 13 pt.

Metric Weight and Mass

1) 60,000 g
2) 24,000 g
3) 610,000 g
4) 82,000 g
5) 95,800g
6) 4,850 g
7) 1,500 g
8) 95 kg
9) 241 kg
10) 700 kg
11) 2.5 kg
12) 28.9 kg
13) 970 kg
14) 325.5 kg

Customary Weight and Mass

1) 5 T
2) 9.5 T
3) 16 T
4) 8.4 T
5) 432 oz
6) 406.4 oz

WWW.MathNotion.Com

GMAS Subject Test Mathematics Grade 5

7) 1,984 oz
8) 8,000 lb.
9) 14,000 lb.
10) 22,400 lb.
11) 25,600 lb.
12) 38,400 oz
13) 292,000 oz
14) 24,000 oz

Temperature

1) −6.11°C
2) −14.72°C
3) 22°C
4) 120°C
5) 24°C
6) 59°C
7) 75°C
8) 102°C
9) 40°C
10) 90°C
11) 95°C
12) 26°C
13) 35.6°F
14) 66.2°F
15) 118.4°F
16) 129.2°F
17) 177.8°F
18) 75.2°F
19) 190.4°F
20) 161.6°F
21) 176°F
22) 60.8°F
23) 50°F
24) 230°F

Time - Convert

1) 1,320 min
2) 650 weeks
3) 18,720 sec
4) 1,260 sec
5) 30 hr
6) 2 year
7) 13,140 hr
8) 1,008 hr
9) 2,880 min
10) 11 hr
11) 120 months
12) 35.4 min
13) 7 days
14) 133 days

Time - Gap

1) 3:10
2) 4:25
3) 4:30 P.M.
4) 3:20
5) 4:20
6) 2:25
7) 4:30
8) 35 minutes
9) 47 minutes

Add Money

1) 456, 470, 336
2) 581, 981, 506
3) 757.5, 687.5, 274.31
4) 497.2, 1210.75, 976.7

Subtract Money

5) 391, 295, 227
6) 437.9, 299.95, 222.6
7) 95.8, 206.3, 643.9
8) $9.35

Money: word problem

1) 5
2) $53.08
3) 16
4) 30.5
5) 24.27
6) 51.84

Chapter 8 : Algebraic Thinking

Topics that you'll learn in this chapter:

- ✓ Finding Rules,
- ✓ Algebraic Word Problems,
- ✓ Evaluating Expressions,
- ✓ Variables and Expressions,

Finding Rules

✎ Complete the output.

1- **Rule:** the output is $x - 10.5$

Input	x	15	18	27	32.25	48.5
Output	y					

17) **Rule:** the output is $x \times 5\frac{1}{3}$

Input	x	3	9	15	21	33
Output	y					

2- **Rule:** the output is $x \div 9$

Input	x	513	387	342	198	126
Output	y					

✎ Find a rule to write an expression.

3- **Rule:** _____

Input	x	4	14	19	24
Output	y	10	35	47.5	60

4- **Rule:** _____

Input	x	5	13	19.6	34.5
Output	y	14.4	22.4	29	43.9

5- **Rule:** _____

Input	x	72	96	132	230.4
Output	y	9	12	16.5	28.8

GMAS Subject Test Mathematics Grade 5

Algebraic Word Problems

Circle the number sentence that fits the problem. Then solve for x.

1) Mary had $42. Then she earned more money (x). Now she has $86.

   $42 + x = $86   OR $42 + $86 = x

   x = ____

2) Lisa had $35. Then she earned more money (x). Now she has $78.

   $35 + x = $78   OR $35 + $78 = x

   x = ____

3) Matthew had $37. Then he earned more money (x). Now he has $98.

   $37 + x = $98   OR $37 + $98 = x

   x = ____

4) Charlotte gave 19 of the cookies he had baked to a friend and now he has 45 cookies left. 45 − 19 = x   OR   x − 19 = 45

   x = ____

5) Mia gave 32 of the cookies she had baked to a friend and now she has 55 cookies left. 55 − 32 = x   OR   x − 32 = 55

   x = ____

6) Lucas gave 41 of the cookies he had baked to a friend and now he has 49 cookies left. . 49 − 41 = x   OR   x − 41 = 49

   x = ____

GMAS Subject Test Mathematics Grade 5

Evaluate Expressions

✍ Simplify each algebraic expression.

1) $11 - x$, $x = 3$

2) $x + 14$, $x = 4$

3) $8 - 3x$, $x = 1$

4) $3x + \frac{1}{3}$, $x = \frac{1}{2}$

5) $2x + 18$, $x = 1.5$

6) $7 - 3x$, $x = 1.2$

7) $12 + 2x - 15$, $x = 3.5$

8) $25 - 5x$, $x = 2.2$

9) $\frac{44}{x} - 58$, $x = 0.4$

10) $\frac{x}{3} - 15 + x$, $x = 12.6$

11) $\frac{x}{7} + 9.5$, $x = 28.7$

12) $\frac{33}{x} - 5.2 + 2.1x$, $x = 3$

13) $2x - \frac{45}{x} - 12$, $x = 9$

14) $\frac{x}{17} - 1.8$, $x = 34$

15) $2(12.5x + 8)$, $x = 2$

16) $16x + 13x - 27 + 9$,

$x = 1.5$

17) $8.7 - \frac{16}{x} + 3x$,

$x = 4$

18) $5(3a - 2a)$,

$a = 5.5$

19) $14 - 2x + 16 - x$,

$x = 3.5$

20) $6x - 3 - x$,

$x = 1.6$

21) $18 - 2(2x + x)$, $x = 0.2$

WWW.MathNotion.Com

GMAS Subject Test Mathematics Grade 5

Variables and Expressions

✎ Write a verbal expression for each algebraic expression.

1) $3a - 7b$

2) $8.2c^2 + 5d$

3) $x - 19.5$

4) $\dfrac{90}{6}$

5) $x^2 + y^3$

6) $4x + 9$

7) $x^2 - 6y + 15$

8) $x^3 + 7y^2 - 6$

9) $\dfrac{1}{5}x + \dfrac{1}{4}y - 11$

10) $\dfrac{1}{7}(x + 13) - 18.6y$

✎ Write an algebraic expression for each verbal expression.

11) 14 less than h

12) The product of 19 and a

13) The 23.2 divided by K

14) The product of 7 and the third power of x

15) 14 more than h to the sixth power

16) 30 more than triple d

17) One eighth the square of b

18) The difference of 42.5 and 3 times a number

19) 73 more than the cube of a number

20) one-quarters the cube of a number

GMAS Subject Test Mathematics Grade 5

Answers of Worksheets

Finding Rules

1)

Input	x	15	18	27	32.25	48.5
Output	y	4.5	7.5	16.5	21.75	38

2)

Input	x	3	9	15	21	33
Output	y	16	48	80	112	176

3)

Input	x	513	387	342	198	126
Output	y	57	43	38	22	14

4) $y = 2.5x$ 5) $y = x + 9.4$ 6) $y = x \div 8$

Algebraic Word Problems

1) $\$42 + x = \$86; x = 44$ 4) $x - 19 = 45; x = 64$

2) $\$35 + x = \$78; x = 43$ 5) $x - 32 = 55; x = 87$

3) $\$37 + x = \$98; x = 61$ 6) $x - 41 = 49; x = 90$

Evaluating Expressions

1) 8   7) 4   13) 1   19) 19.5

2) 18   8) 14   14) 0.2   20) 5

3) 5   9) 52   15) 66   21) 16.8

4) $\frac{11}{6}$ 10) 1.8   16) 25.5

5) 21   11) 13.6   17) 16.7

6) 3.4   12) 12.1   18) 27.5

Variables and Expressions

1) 3 times a minus 7 times b.

2) 8.2 times c squared plus 5 times d.

3) a number minus 19.5.

4) the quotient of 90 and 6.

5) x squared plus y cubed.

6) the product of 4 and x plus 9.

GMAS Subject Test Mathematics Grade 5

7) x squared plus the product of 6 and y plus 15.

8) x cubed plus the product of 7 and y squared minus the product of 6 and y.

9) the sum of one–fifth of x and one–quarters of y, minus 11.

10) one–seventh of the sum of x and 13 minus the product of 18.6 and y.

11) $14 < h$

12) $19a$

13) $\frac{23.2}{K}$

14) $7x^3$

15) $14 > h^6$

16) $3d < 30$

17) $\frac{1}{8}b^2$

18) $42.5 - 3a$

19) $73 > a^3$

20) $\frac{1}{4}x^3$

Chapter 9 : Geometric

Topics that you'll learn in this chapter:

- ✓ Identifying Angles,
- ✓ Estimate and Measure Angles,
- ✓ Polygon Names,
- ✓ Classify Triangles,
- ✓ Parallel Sides in Quadrilaterals,
- ✓ Identify Parallelograms,
- ✓ Identify Trapezoids,
- ✓ Identify Rectangles,
- ✓ Perimeter and Area of Squares,
- ✓ Perimeter and Area of rectangles,
- ✓ Area and Perimeter: Word Problems,
- ✓ Circumference, Diameter and Radius,
- ✓ Volume of Cubes and Rectangle Prisms,

GMAS Subject Test Mathematics Grade 5

Identifying Angles

✏ Write the name of the angles (Acute, Right, Obtuse, and Straight).

1)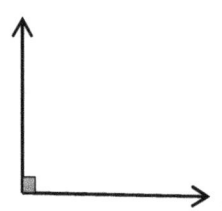

2)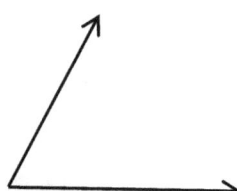

3)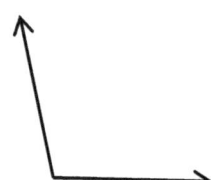

4)

5)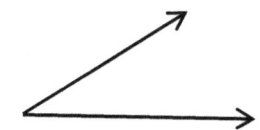

6)

7)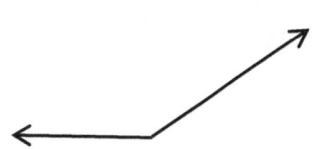

8)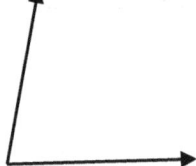

GMAS Subject Test Mathematics Grade 5

Estimate Angle Measurements

🖊 Estimate the approximate measurement of each angle in degrees.

1)

2)

3)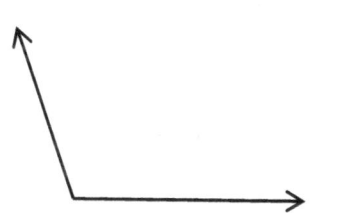

4)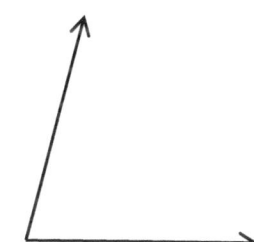

5)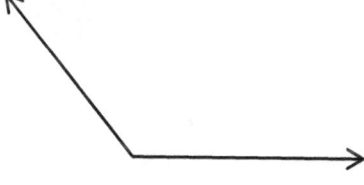

6)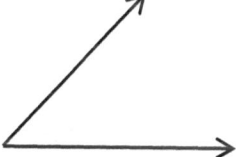

7)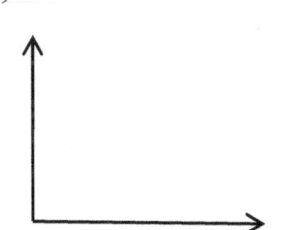

8)

WWW.MathNotion.Com

GMAS Subject Test Mathematics Grade 5

Measure Angles with a Protractor

✏️ Use protractor to measure the angles below.

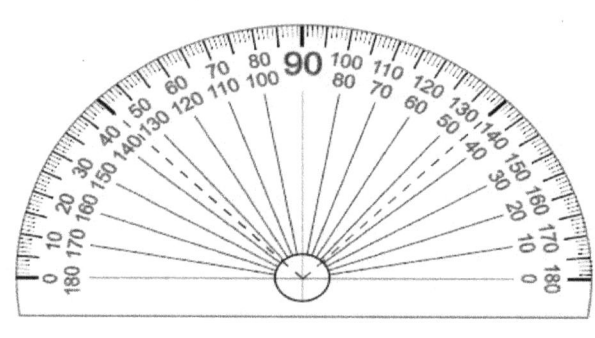

1)

2)

3)

4)

✏️ Use a protractor to draw angles for each measurement given.

1) 140°

2) 100°

3) 110°

4) 120°

5) 55°

Polygon Names

Write name of polygons.

1)

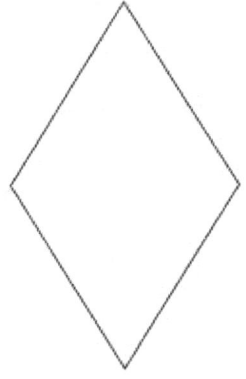

2)

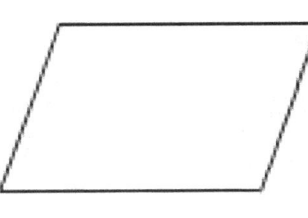

3)

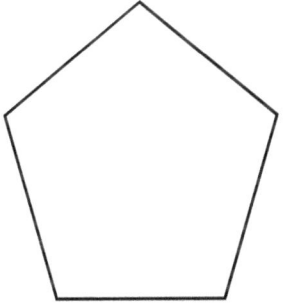

4)

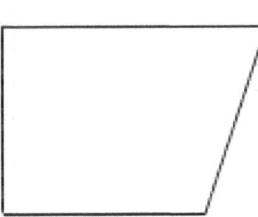

5)

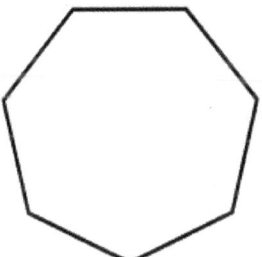

6)

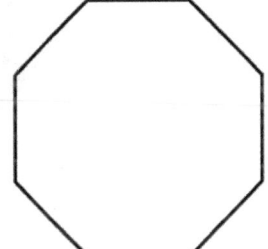

Classify Triangles

✎ Classify the triangles by their sides and angles.

1)

2)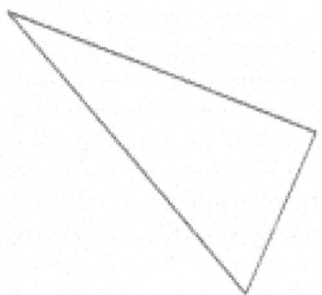

3)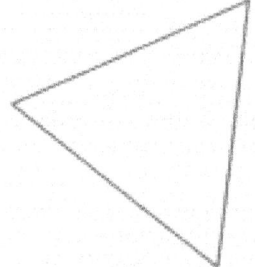

4)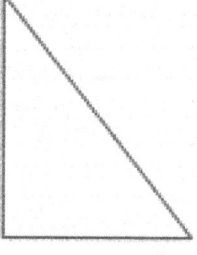

5)

6)

Parallel Sides in Quadrilaterals

✎ Write name of quadrilaterals.

1)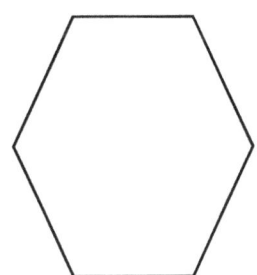

2)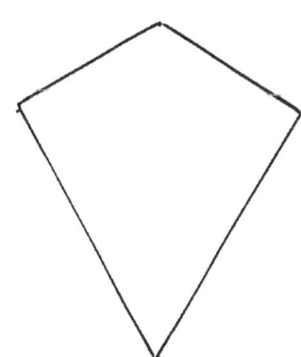

3)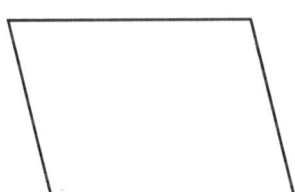

4)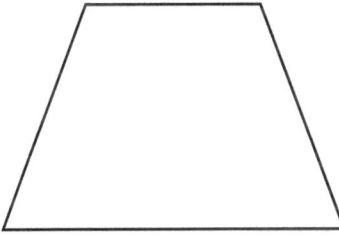

5)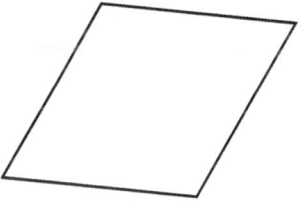

6)

Identify Rectangles

✏ Solve.

1) A rectangle has _____ sides and _____ angles.

2) Draw a rectangle that is 5.5 centimeters long and 2.5 centimeters wide. What is the perimeter?

3) Draw a rectangle 3.5 cm long and 1.5 cm wide.

4) Draw a rectangle whose length is 4.25 cm and whose width is 2.45 cm. What is the perimeter of the rectangle?

5) What is the perimeter of the rectangle?

7.2

5.8

GMAS Subject Test Mathematics Grade 5

Perimeter: Find the Missing Side Lengths

✏ Find the missing side of each shape.

1) perimeter = 57.2

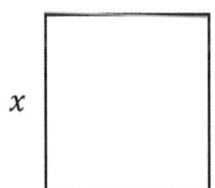

2) perimeter = 21.2

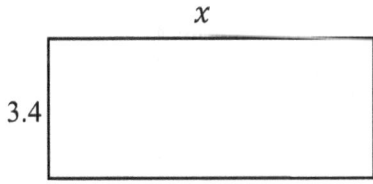

3) perimeter = 27.5

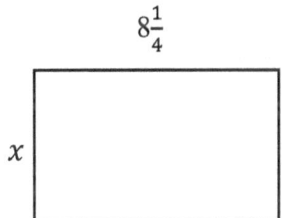

4) perimeter = 35.2

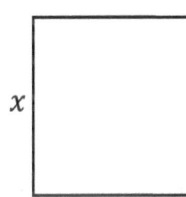

5) perimeter = 75.6

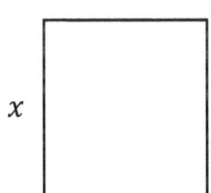

6) perimeter = 30.8

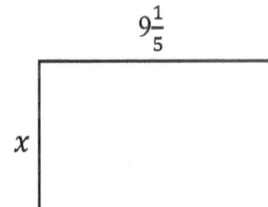

7) perimeter = 36.25

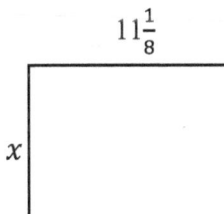

8) perimeter = 46.8

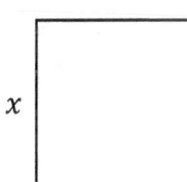

WWW.MathNotion.Com

GMAS Subject Test Mathematics Grade 5

Perimeter and Area of Squares

✏️ Find perimeter and area of squares.

1) A: _____ , P: _____

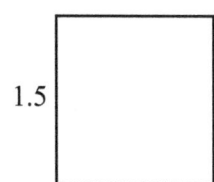

2) A: _____ , P: _____

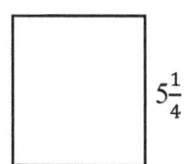

3) A: _____ , P: _____

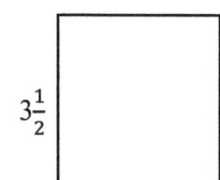

4) A: _____ , P: _____

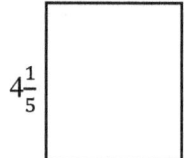

5) A: _____ , P: _____

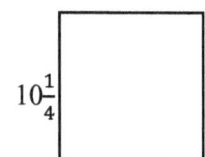

6) A: _____ , P: _____

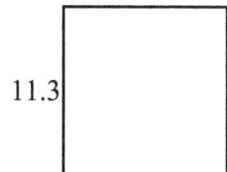

7) A: _____ , P: _____

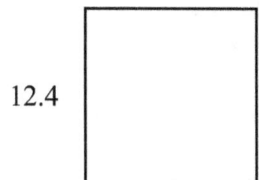

8) A: _____ , P: _____

GMAS Subject Test Mathematics Grade 5

Perimeter and Area of rectangles

✎ Find perimeter and area of rectangles.

1) A: _____ , P: _____

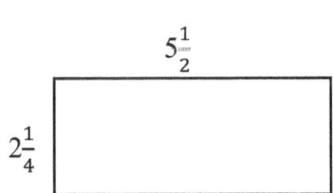

2) A: _____ , P: _____

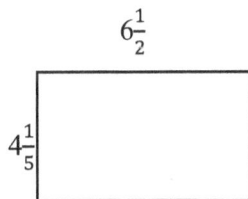

3) A: _____ , P: _____

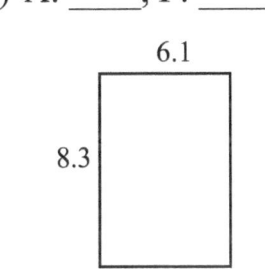

4) A: _____ , P: _____

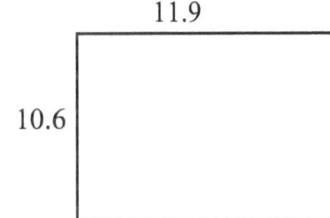

5) A: _____ , P: _____

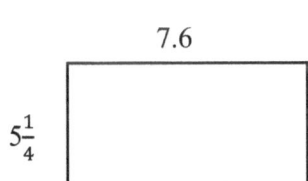

6) A: _____ , P: _____

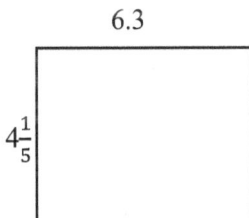

7) A: _____ , P: _____

9.7
$3\frac{1}{2}$

8) A: _____ , P: _____

11.4
5.8

GMAS Subject Test Mathematics Grade 5

Find the Area or Missing Side Length of a Rectangle

✏ Find area or missing side length of rectangles.

1) Area =?

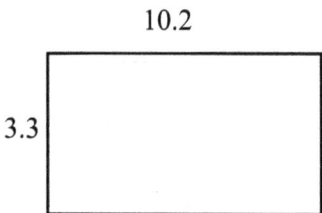

2) Area = 42.12, x=?

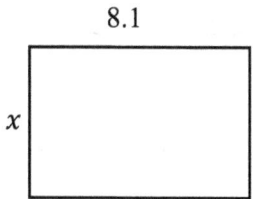

3) Area = 29.52, x=?

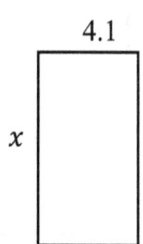

4) Area =?

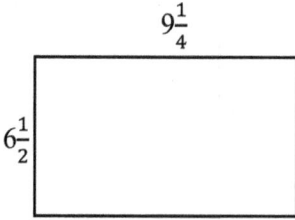

5) Area =?

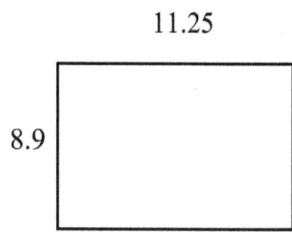

6) Area = 662.34 x=?

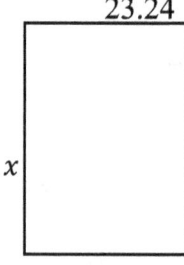

7) Area = 216.24, x=?

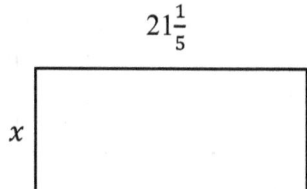

8) Area 336.42, x=?

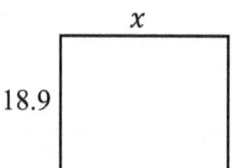

WWW.MathNotion.Com

GMAS Subject Test Mathematics Grade 5

Area and Perimeter: Word Problems

✍ Solve.

1) The area of a rectangle is 90.86 square meters. The width is 7.7 meters. What is the length of the rectangle?

2) A square has an area of 6.25 square feet. What is the perimeter of the square?

3) Ava built a rectangular vegetable garden that is 3.2 feet long and has an area of 21.12 square feet. What is the perimeter of Ava's vegetable garden?

4) A square has a perimeter of 12.8 millimeters. What is the area of the square?

5) The perimeter of David's square backyard is 0.96 meters. What is the area of David's backyard?

6) The area of a rectangle is 37.63 square inches. The length is 7.1 inches. What is the perimeter of the rectangle?

GMAS Subject Test Mathematics Grade 5

Circumference, Diameter, and Radius

✎ Find the diameter and circumference of circles.

1)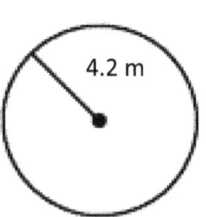

2)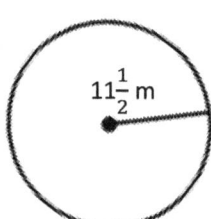

3)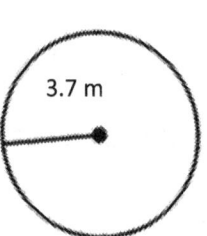

4)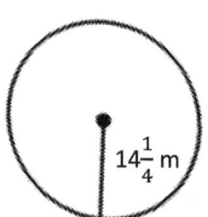

✎ Find the radius.

5)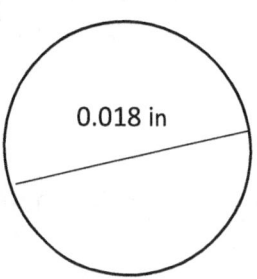

6)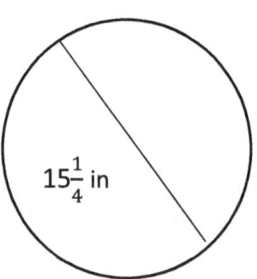

7) Diameter = $16\frac{1}{5} ft$

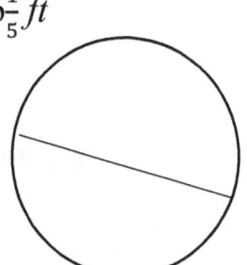

8) Diameter = 23.82 m

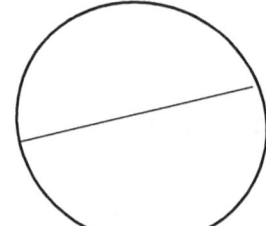

GMAS Subject Test Mathematics Grade 5

Volume of Cubes and Rectangle Prisms

🖎 Find the volume of each of the rectangular prisms.

1)

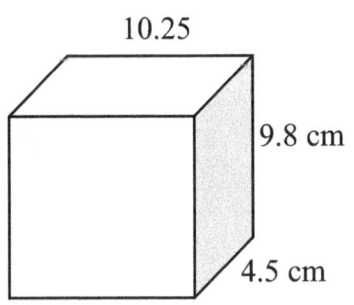

2)

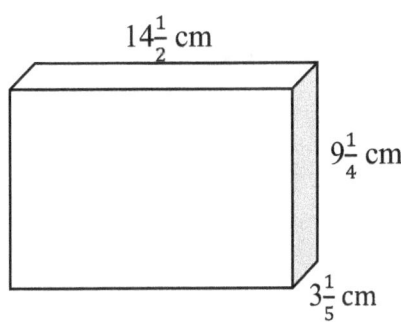

3)

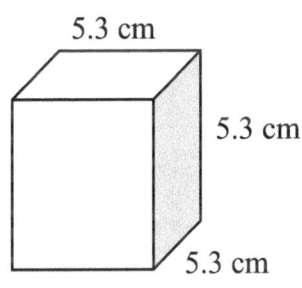

4)

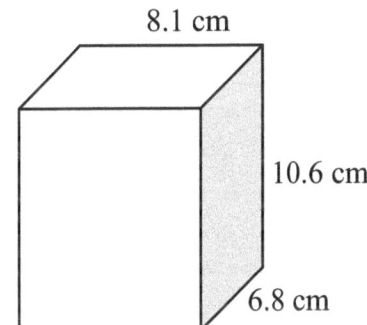

5)

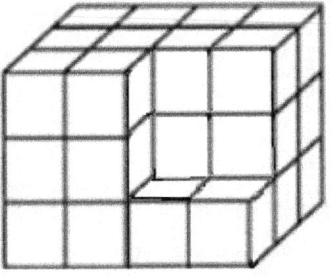

6)

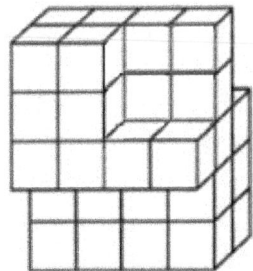

WWW.MathNotion.Com   103

GMAS Subject Test Mathematics Grade 5

Answers of Worksheets

Identifying Angles

1) Right 3) Obtuse 5) Acute 7) Obtuse
2) Acute 4) Straight 6) Obtuse 8) Acute

Estimate Angle Measurements

1) 160° 3) 110° 5) 130° 7) 90°
2) 180° 4) 75° 6) 45° 8) 60°

Measure Angles with a Protractor

1) 50° 2) 135° 3) 20° 4) 170°

Draw Angles

1) 2) 3)

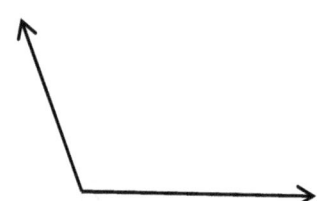

4) 5)

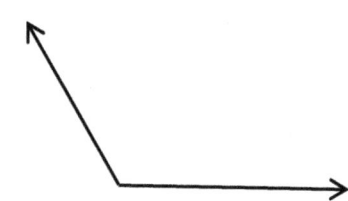 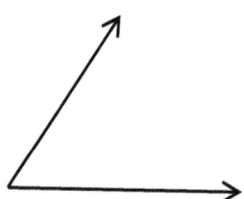

Polygon Names

1) Diamond 3) Pentagon 5) Heptagon
2) Parallelogram 4) Trapezius 6) Octagon

Classify Triangles

1) Scalene, acute   4) Scalene, right
2) Isosceles, acute   5) Isosceles, right
3) Equilateral, acute   6) Scalene, obtuse

GMAS Subject Test Mathematics Grade 5

Parallel Sides in Quadrilaterals

1) Hexagon
2) Kike
3) Parallelogram
4) Trapezoid
5) Rhombus
6) Rectangle

Identify Rectangles

1) 4 - 4
2) 16
3) Draw the rectangle.
4) 13.4
5) 26

Perimeter: Find the Missing Side Lengths

1) 14.3
2) 7.2
3) 5.5
4) 8.8
5) 18.9
6) 6.2
7) 7
8) 11.7

Perimeter and Area of Squares

1) A: 2.25, P: 6
2) A: 27.56, P: 21
3) A: 12.25, P: 14
4) A: 17.64, P: 16.8
5) A: 105.063 P: 41
6) A: 127.69, P: 45.2
7) A: 153.76, P: 49.6
8) A: 96.04, P: 39.2

Perimeter and Area of rectangles

1) A: 12.375, P: 15.5
2) A: 27.3, P: 21.4
3) A: 50.63, P: 28.8
4) A: 126.14, P: 45
5) A: 39.9, P: 25.7
6) A: 26.46, P: 21
7) A: 33.95, P: 26.4
8) A: 66.12, P: 34.4

Find the Area or Missing Side Length of a Rectangle

1) 33.66
2) 5.2
3) 7.2
4) 60.125
5) 100.125
6) 28.5
7) 10.2
8) 17.8

Area and Perimeter: Word Problems

1) 11.8
2) 10
3) 19.6
4) 10.24
5) 0.0576
6) 24.8

Circumference, Diameter, and Radius

1) diameter: 8.4 circumferences: 8.4π or 26.376
2) diameter: 23 circumferences: 23π or 72.22
3) diameter: 7.4 circumferences: 7.4π or 23.24
4) diameter: 28.5 circumferences: 28.5π or 89.49
5) radius: 0.009 in
6) radius: 7.625 in
7) radius: 8.1 ft
8) radius: 11.91 m

Volume of Cubes and Rectangle Prisms

1) 452.025 cm^3
2) 429.2 cm^3
3) 148.877 c m^3
4) 583.848 cm^3
5) 32
6) 40

WWW.MathNotion.Com

GMAS Subject Test Mathematics Grade 5

Chapter 10 : Three-Dimensional Figures

Topics that you'll learn in this chapter:

- ✓ Identify Three–Dimensional Figures,
- ✓ Count Vertices, Edges, and Faces,
- ✓ Identify Faces of Three–Dimensional Figures,

Identify Three–Dimensional Figures

✎ Write the name of each shape.

1)

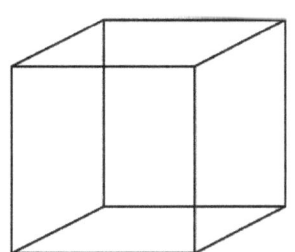

2)

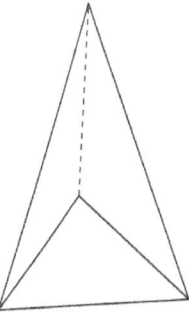

3)

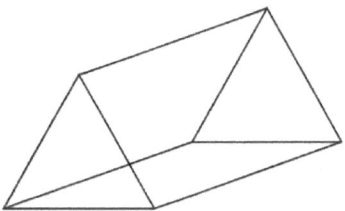

4)

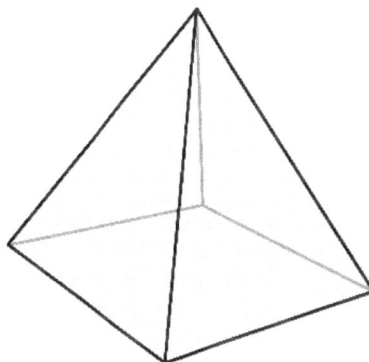

5)

6)

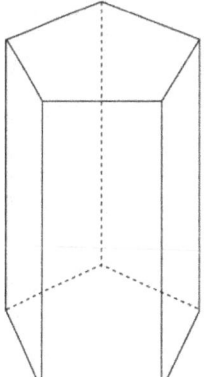

Count Vertices, Edges, and Faces

	Shape	Number of edges	Number of faces	Number of vertices
1)		___	___	___
2)		___	___	___
3)		___	___	___
4)		___	___	___
5)		___	___	___
6)		___	___	___

Identify Faces of Three–Dimensional Figures

✍ Write the number of faces.

1)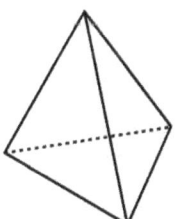

2)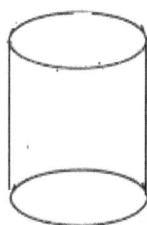

3)

4)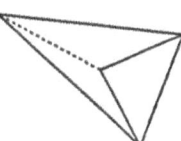

5)

6)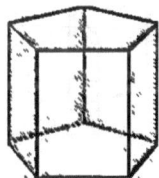

7)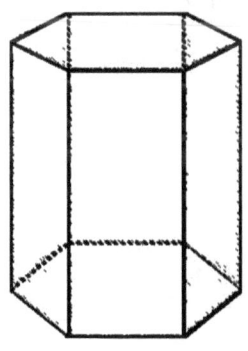

8)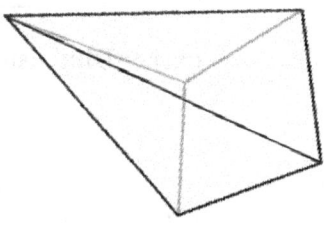

GMAS Subject Test Mathematics Grade 5

Answers of Worksheets

Identify Three–Dimensional Figures

1) Cube

2) Triangular pyramid

3) Triangular prism

4) Square pyramid

5) Rectangular prism

6) Pentagonal prism

7) Hexagonal prism

Count Vertices, Edges, and Faces

Shape	Number of edges	Number of faces	Number of vertices
1)	6	4	4
2)	8	5	5
3)	12	6	8
4)	15	7	10
5)	12	6	8
6)	18	8	12

Identify Faces of Three–Dimensional Figures

1) 6

2) 2

3) 5

4) 4

5) 6

6) 7

7) 8

8) 5

WWW.MathNotion.Com

Chapter 11 : Symmetry and Transformations

Topics that you'll learn in this chapter:

- ✓ Line Segments,
- ✓ Identify Lines of Symmetry,
- ✓ Count Lines of Symmetry,
- ✓ Parallel, Perpendicular and Intersecting Lines,

Line Segments

✎ Write each as a line, ray, or line segment.

1)

2)

3)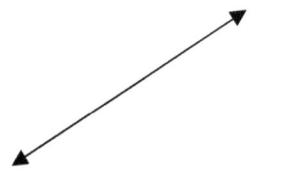

4)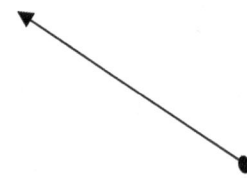

5)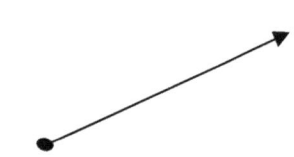

6)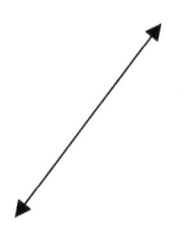

7)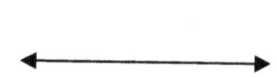

8)

Identify Lines of Symmetry

 Tell whether the line on each shape a line of symmetry is.

1)
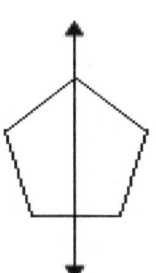

2)
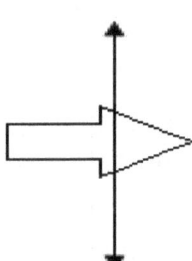

3)
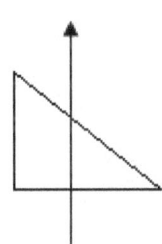

4)

5)
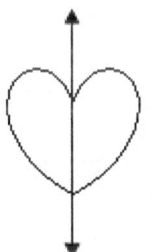

6)
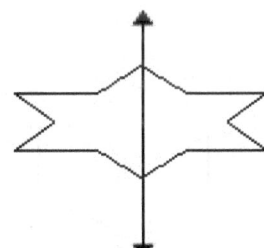

7)
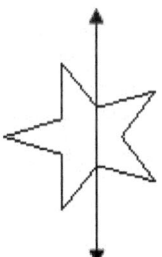

8)
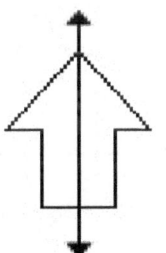

Count Lines of Symmetry

✎ Draw lines of symmetry on each shape. Count and write the lines of symmetry you see.

1)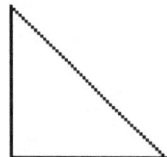

2)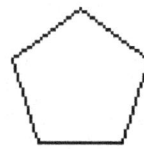

3)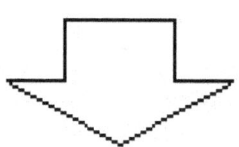

4)

5)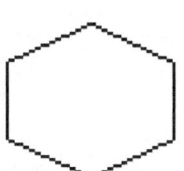

6)

7)

8)

Parallel, Perpendicular and Intersecting Lines

🖎 State whether the given pair of lines are parallel, perpendicular, or intersecting.

1)
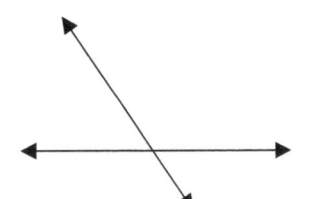

2)

3)
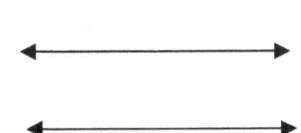

4)
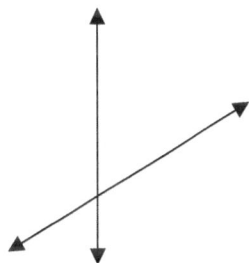

5)
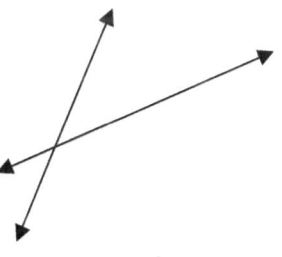

6)
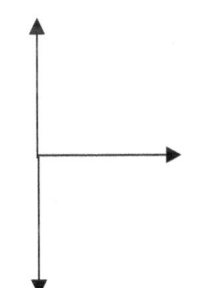

7)
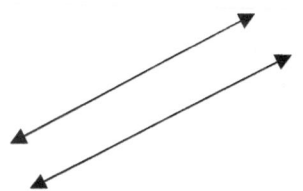

8)
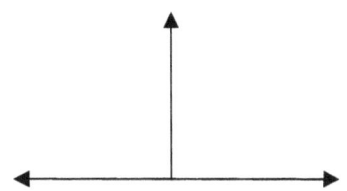

GMAS Subject Test Mathematics Grade 5

Answers of Worksheets

Line Segments

1) Ray
2) Line segment
3) Line
4) Ray
5) Ray
6) Line
7) Line
8) Line segment

Identify lines of symmetry

1) yes
2) no
3) no
4) yes
5) yes
6) yes
7) no
8) yes

Count lines of symmetry

1) 2) 3) 4)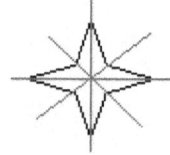

5) 6) 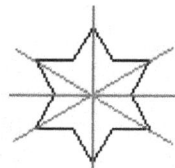 7) 8)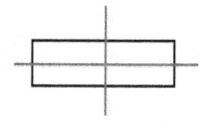

Parallel, Perpendicular and Intersecting Lines

1) Intersection
2) Perpendicular
3) Parallel
4) Intersection
5) Intersection
6) Perpendicular
7) Parallel
8) Perpendicular

Chapter 12 : Data Graphs, and Statistics

Topics that you'll learn in this chapter:

- ✓ Mean, Median, Mode, and Range,
- ✓ Graph Points on a Coordinate Plane,
- ✓ Bar Graph,
- ✓ Tally and Pictographs,
- ✓ Dot Plots
- ✓ Line Graphs,
- ✓ Stem-And-Leaf Plot,
- ✓ Scatter Plots,
- ✓ Probability Problems,

GMAS Subject Test Mathematics Grade 5

Mean and Median

✎ Find Mean and Median of the Given Data.

1) 13, 22, 12, 6, 14

2) 7, 18, 11, 15, 2, 23

3) 33, 25, 14, 7, 14

4) 6, 7, 5, 1, 2, 5

5) 11, 4, 15, 7, 8, 17, 22

6) 9, 2, 5, 2, 18, 14, 45

7) 19, 14, 20, 12, 34, 14, 17, 31

8) 38, 29, 5, 3, 14, 9, 32

9) 28, 27, 31, 36, 32, 53, 41

10) 12, 17, 6, 17, 2, 17, 9, 12

11) 35, 11, 25, 54, 42, 41

12) 42, 48, 50, 48, 27, 67

13) 75, 72, 30, 46, 38, 29

14) 89, 73, 67, 46, 54, 84, 34

15) 78, 21, 100, 85, 54, 60

16) 41, 65, 9, 88, 17, 38, 14

✎ Solve.

17) In a javelin throw competition, five athletics score 54, 72, 59, 67 and 85 meters. What are their Mean and Median? _____

18) Eva went to shop and bought 18 apples, 7 peaches, 6 bananas, 3 pineapple and 9 melons. What are the Mean and Median of her purchase?

_____

WWW.MathNotion.Com

GMAS Subject Test Mathematics Grade 5

Mode and Range

✏️ Find Mode and Rage of the Given Data.

1) 12, 4, 8, 2, 9, 4
   Mode: _____ Range: _____

2) 8, 8, 11, 8, 18, 2, 5, 41
   Mode: _____ Range: _____

3) 3, 3, 2, 23, 4, 12, 3, 9, 3
   Mode: _____ Range: _____

4) 12, 39, 5, 25, 4, 4, 27, 8
   Mode: _____ Range: _____

5) 5, 5, 8, 5, 14, 3, 12
   Mode: _____ Range: _____

6) 2, 5, 19, 15, 12, 11, 7, 8, 7, 7
   Mode: _____ Range: _____

7) 9, 4, 0, 12, 19, 21, 9, 7, 9, 3
   Mode: _____ Range: _____

8) 11, 4, 3, 11, 5, 8, 44, 11, 7
   Mode: _____ Range: _____

9) 3, 3, 6, 9, 3, 3, 9, 15, 14, 20
   Mode: _____ Range: _____

10) 15, 14, 14, 16, 20, 7, 1, 30
    Mode: _____ Range: _____

11) 13, 3, 27, 4, 4, 16, 18, 4
    Mode: _____ Range: _____

12) 7, 22, 35, 11, 7, 24, 6, 13
    Mode: _____ Range: _____

✏️ Solve.

13) A stationery sold 14 pencils, 28 red pens, 51 blue pens, 14 notebooks, 27 erasers, 43 rulers and 40 color pencils. What are the Mode and Range for the stationery sells?

    Mode: _____   Range: _____

14) In an English test, eight students score 14, 20, 22, 14, 17, 28, 36 and 24. What are their Mode and Range? _____

WWW.MathNotion.Com

Graph Points on a Coordinate Plane

🖉 Plot each point on the coordinate grid.

1) A (4, 6) 3) C (1, 5) 5) E (4, 8)
2) B (3, 2) 4) D (5, 7) 6) F (9, 2)

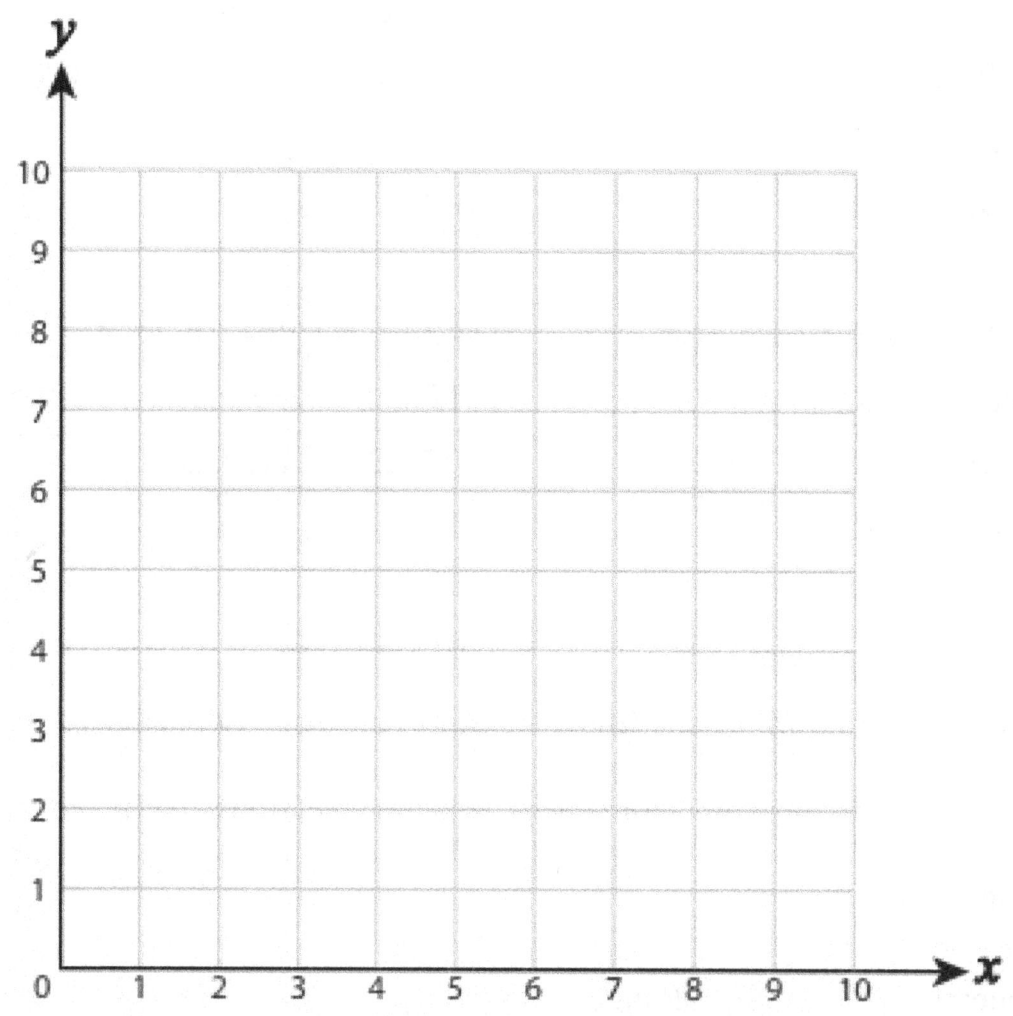

Bar Graph

1) Graph the given information as a bar graph.

Day	Hot dogs sold
Monday	50
Tuesday	80
Wednesday	10
Thursday	30
Friday	70

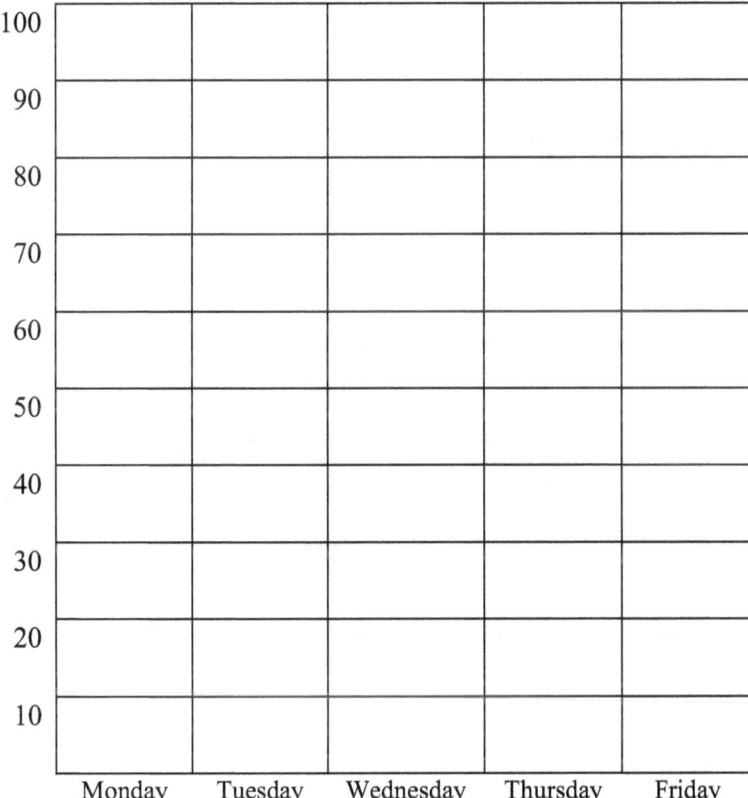

Tally and Pictographs

Using the key, draw the pictograph to show the information.

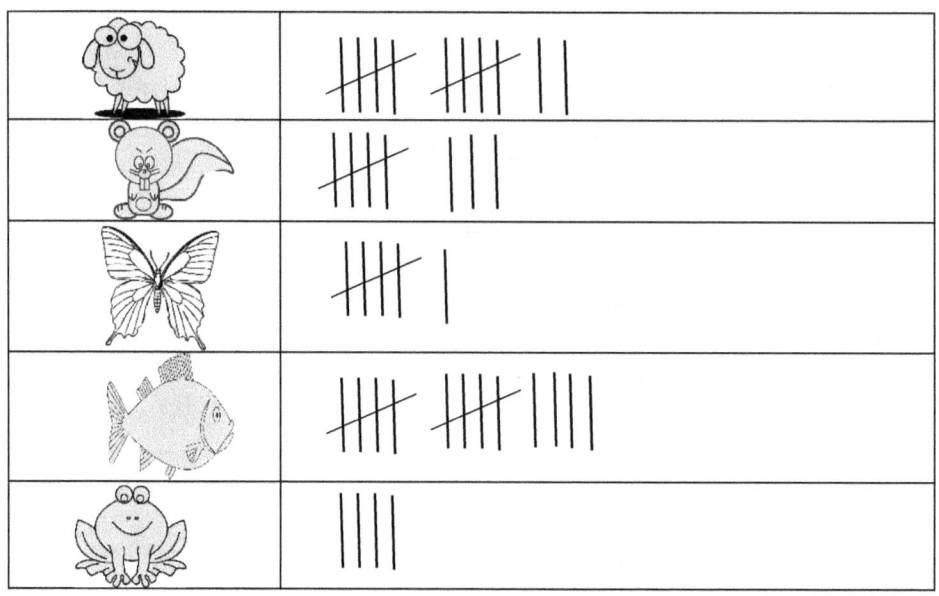

Key: = 2 animals

Dot plots

The ages of students in a Math class are given below.

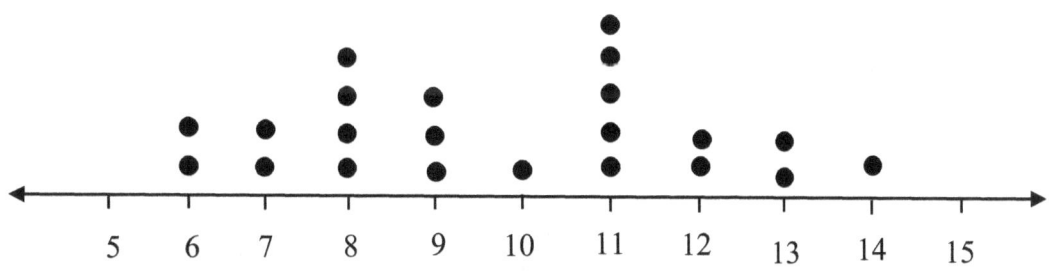

1) What is the total number of students in math class?

2) How many students are at least 12 years old?

3) Which age(s) has the most students?

4) Which age(s) has the fewest student?

5) Determine the median of the data.

6) Determine the range of the data.

7) Determine the mode of the data.

Line Graphs

David work as a salesman in a store. He records the number of shoes sold in five days on a line graph. Use the graph to answer the question

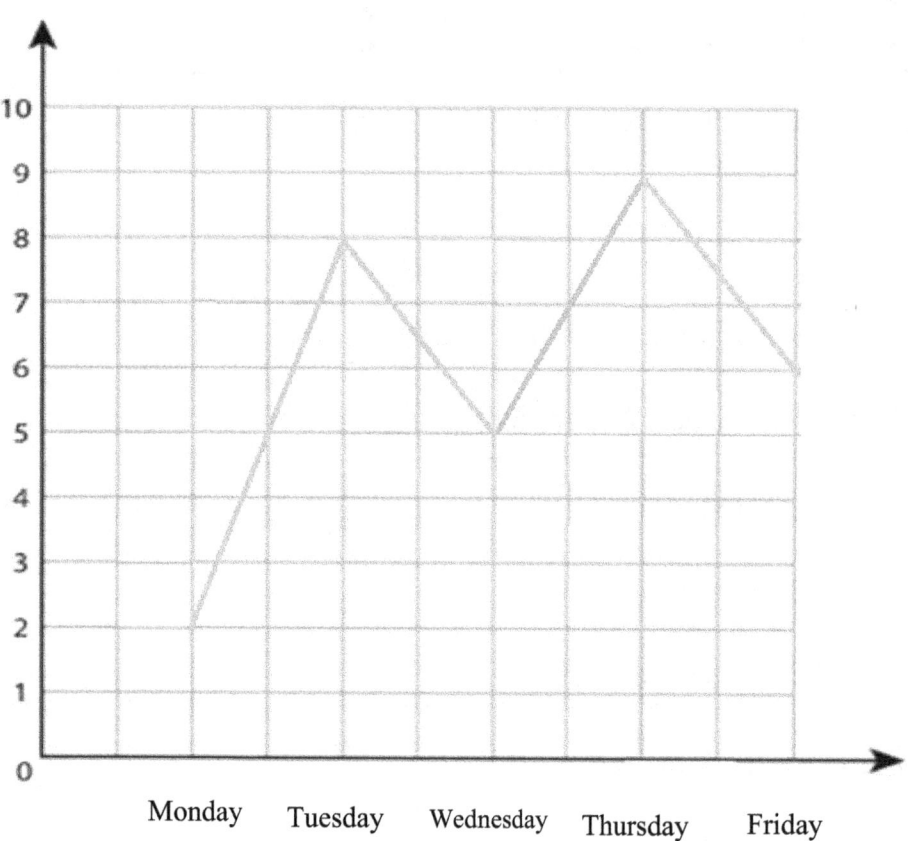

1) How many shoes were sold on Friday?

2) Which day had the minimum sales of shoes?

3) Which day had the maximum number of shoes sold?

4) How many shoes were sold in 5 days?

GMAS Subject Test Mathematics Grade 5

Stem–And–Leaf Plot

✎ Make stem ad leaf plots for the given data.

1) 42, 47, 14, 19, 42, 69, 65, 49, 42, 10, 64

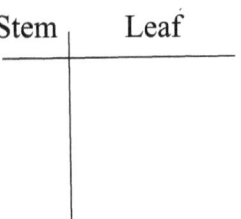

2) 43, 85, 52, 48, 45, 43, 51, 81, 59, 50, 85, 89

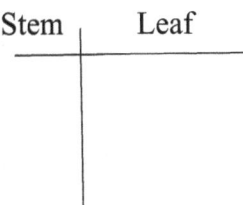

3) 112, 39, 46, 35, 80, 119, 42, 114, 37, 112, 47, 119

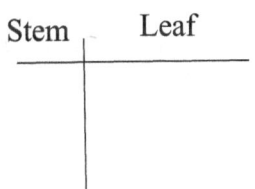

4) 90, 50, 131, 93, 112, 56, 139, 98, 115, 59, 98, 135, 111

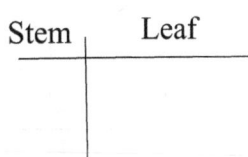

Scatter Plots

🖊 Construct a scatter plot.

x	1	2	3	4	5	8
y	15	40	50	35	70	20

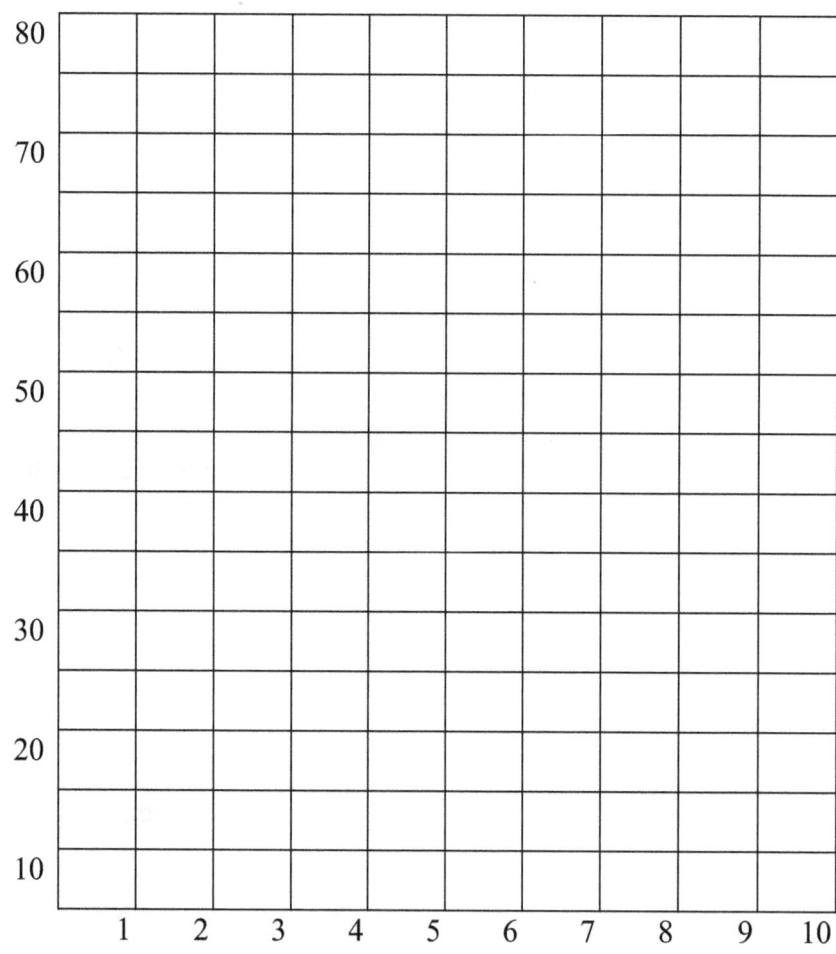

Probability Problems

✏️ Solve.

1) A number is chosen at random from 1 to 20. Find the probability of selecting a 10 or smaller.

2) A number is chosen at random from 1 to 25. Find the probability of selecting multiples of 5.

3) A number is chosen at random from 1 to 15. Find the probability of selecting multiples of 7.

4) A number is chosen at random from 1 to 20. Find the probability of selecting a multiple of 6.

5) A number is chosen at random from 1 to 10. Find the probability of selecting prime numbers.

6) A number is chosen at random from 1 to 20. Find the probability of not selecting factors of 15.

GMAS Subject Test Mathematics Grade 5

Answers of Worksheets

Mean and Median

1) Mean: 13.4, Median: 13
2) Mean: 12.67, Median: 13
3) Mean: 18.6, Median: 14
4) Mean: 4.33, Median: 5
5) Mean: 12, Median: 11
6) Mean: 13.57, Median: 9
7) Mean: 20.125, Median: 18
8) Mean: 18.57, Median: 14
9) Mean: 35.43, Median: 32
10) Mean: 11.5, Median: 12
11) Mean: 34.67, Median: 38
12) Mean: 47, Median: 48
13) Mean: 48.33, Median: 42
14) Mean: 63.86, Median: 67
15) Mean: 66.33, Median: 69
16) Mean: 38.86, Median: 38
17) Mean: 67.4, Median: 67
18) Mean: 8.6, Median: 7

Mode and Range

1) Mode: 4, Range: 10
2) Mode: 8, Range: 39
3) Mode: 3, Range: 21
4) Mode: 4, Range: 35
5) Mode: 5, Range: 11
6) Mode: 7, Range: 17
7) Mode: 9, Range: 21
8) Mode: 11, Range: 41
9) Mode: 3, Range: 17
10) Mode: 14, Range: 29
11) Mode: 4, Range: 24
12) Mode: 7, Range: 29
13) Mode: 14, Range: 37
14) Mode: 14, Range: 22

Graph Points on a Coordinate Plane

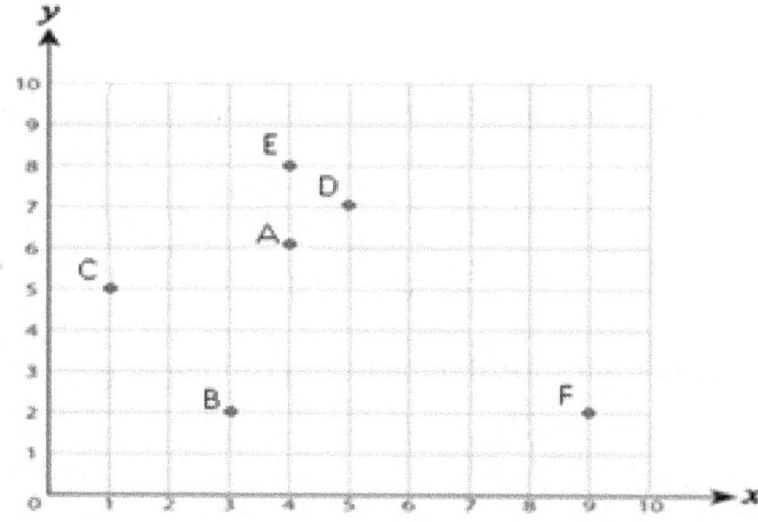

GMAS Subject Test Mathematics Grade 5

Bar Graph

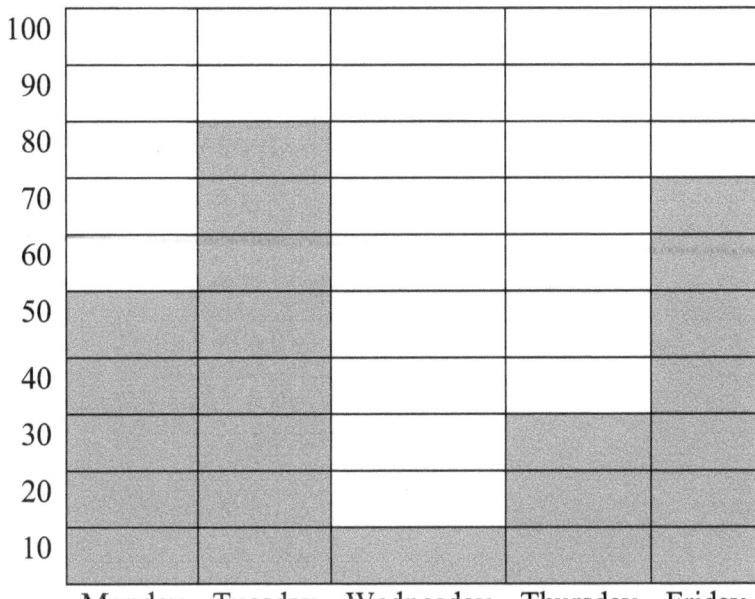

Tally and Pictographs

Dot plots

1) 22
2) 5
3) 11
4) 10 and 14
5) 2
6) 4
7) 2

Line Graphs

1) 6
2) Monday
3) Thursday
4) 30

GMAS Subject Test Mathematics Grade 5

Stem–And–Leaf Plot

1)
Stem	leaf
1	0 4 9
4	2 2 2 7 9
6	4 5 9

2)
Stem	leaf
4	3 3 5 8
5	0 1 2 9
8	1 5 5 9

3)
Stem	leaf
3	5 7 9
4	2 6 7
8	0
11	2 2 4 9 9

4)
Stem	leaf
5	0 6 9
9	0 3 8 8
11	1 2 5
13	1 5 9

Scatter Plots

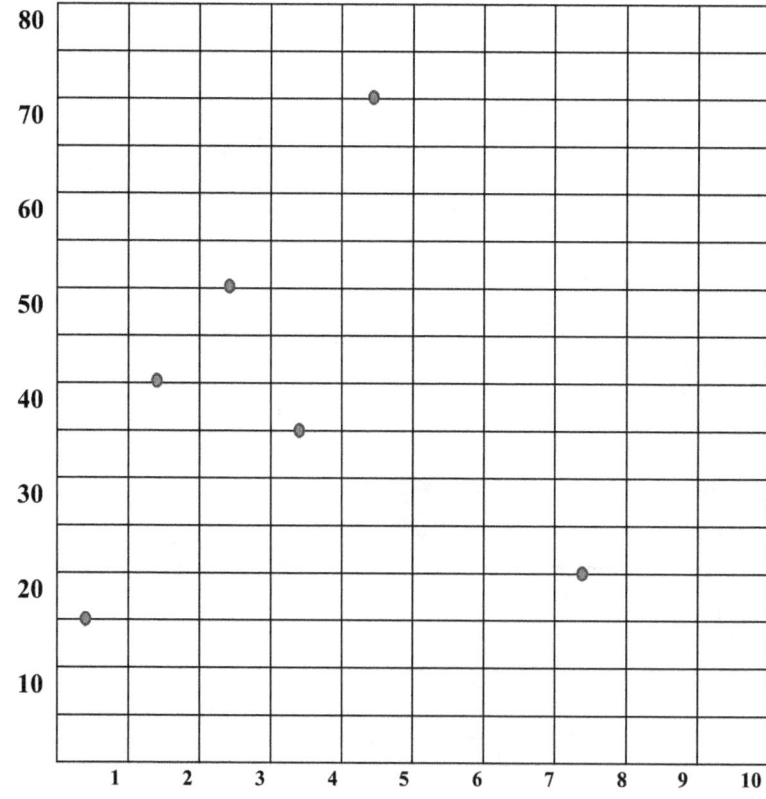

Probability Problems

1) $\frac{1}{2}$

2) $\frac{1}{5}$

3) $\frac{2}{15}$

4) $\frac{3}{20}$

5) $\frac{2}{5}$

6) $\frac{4}{5}$

GMAS Subject Test Mathematics Grade 5

Chapter 13 : GMAS Math Practice Tests

Time to Test

Time to refine your skill with a practice examination.

Take a REAL GMAS Mathematics test to simulate the test day experience. After you've finished, score your test using the answer key.

Before You Start

- You'll need a pencil and scratch papers to take the test.
- For this practice test, don't time yourself. Spend time as much as you need.
- It's okay to guess. You won't lose any points if you're wrong.
- After you've finished the test, review the answer key to see where you went wrong.

Calculators are not permitted for Grade 5 GMAS Tests

Good Luck!

GMAS Subject Test Mathematics Grade 5

GMAS GRADE 5 MAHEMATICS REFRENCE MATERIALS

Perimeter

Square $\qquad$ $P = 4S$

Rectangle $\qquad$ $P = 2L + 2W$

Area

Square $\qquad$ $A = S \times S$

Rectangle $\qquad$ $A = l \times w$ or $A = bh$

Volume

Square $\qquad$ $A = S \times S \times S$

Rectangle $\qquad$ $A = l \times w \times h$ or $A = Bh$

LENGTH

Customary	Metric
1 mile (mi) = 1,760 yards (yd)	1 kilometer (km) = 1,000 meters (m)
1 yard (yd) = 3 feet (ft)	1 meter (m) = 100 centimeters (cm)
1 foot (ft) = 12 inches (in.)	1 centimeter (cm) = 10 millimeters (mm)

VOLUME AND CAPACITY

Customary	Metric
1 gallon (gal) = 4 quarts (qt)	1 liter (L) = 1,000 milliliters (mL)
1 quart (qt) = 2 pints (pt.)	
1 pint (pt.) = 2 cups (c)	
1 cup (c) = 8 fluid ounces (Fl oz)	

WEIGHT AND MASS

Customary	Metric
1 ton (T) = 2,000 pounds (lb.)	1 kilogram (kg) = 1,000 grams (g)
1 pound (lb.) = 16 ounces (oz)	1 gram (g) = 1,000 milligrams (mg)

Georgia Milestones Assessment System Practice Test 1

Mathematics

GRADE 5

Released

GMAS Subject Test Mathematics Grade 5

Session 1

❖ **Calculators are NOT permitted for this practice test.**

❖ **Time for Session 1: 85 Minutes**

GMAS Subject Test Mathematics Grade 5

1) A baker uses 6 eggs to bake a cake. How many cakes will he be able to bake with 240 eggs?

   A. 30

   B. 40

   C. 160

   D. 80

2) The area of a rectangle is D square feet and its length is 9 feet. Which equation represents W, the width of the rectangle in feet?

   A. $W = \dfrac{D}{9}$

   B. $W = \dfrac{9}{D}$

   C. $W = 9D$

   D. $W = 9 + D$

3) Which list shows the fractions in order from least to greatest?

   $$\frac{5}{8}, \frac{4}{9}, \frac{8}{10}, \frac{10}{11}, \frac{3}{13}$$

   A. $\dfrac{5}{8}, \dfrac{10}{11}, \dfrac{4}{9}, \dfrac{3}{13}, \dfrac{8}{10}$

   B. $\dfrac{8}{10}, \dfrac{3}{13}, \dfrac{5}{8}, \dfrac{10}{11}, \dfrac{4}{9}$

   C. $\dfrac{4}{9}, \dfrac{5}{8}, \dfrac{10}{11}, \dfrac{3}{13}, \dfrac{8}{10}$

   D. $\dfrac{3}{13}, \dfrac{4}{9}, \dfrac{5}{8}, \dfrac{8}{10}, \dfrac{10}{11}$

4) If A = 25, then which of the following equations are correct?

   A. A + 25 = 50

   B. A ÷ 25 = 50

   C. 25 × A = 50

   D. A − 25 = 50

5) Which statement about 4 multiplied by $\frac{6}{5}$ is true?

   A. The product is between 5 and 7.

   B. The product is between 4 and 5.

   C. The product is more than $\frac{32}{6}$.

   D. The product is between $\frac{18}{8}$ and 4.

6) The area of a circle is 64π. What is the circumference of the circle?

   Write your answer in the box below.

   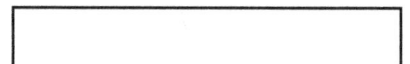

7) What is the volume of this box?

   A. 21 cm³

   B. 28 cm³

   C. 56 cm³

   D. 84 cm³

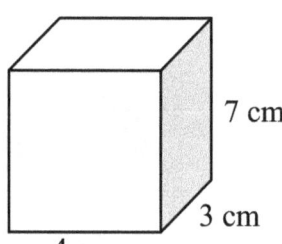

GMAS Subject Test Mathematics Grade 5

8) A shirt costing $180 is discounted 30%. Which of the following expressions can be used to find the selling price of the shirt?

A. (180) (0.30)

B. (180) – 180 (0.70)

C. (180) (0.30) – (180) (0.30)

D. (180) (0.70)

9) In a bag, there are 50 cards. Of these cards, 4 cards are white. What fraction of the cards are white?

A. $\frac{2}{25}$

B. $\frac{4}{25}$

C. $\frac{1}{10}$

D. $\frac{3}{50}$

10) The perimeter of the trapezoid below is 38. What is its area?

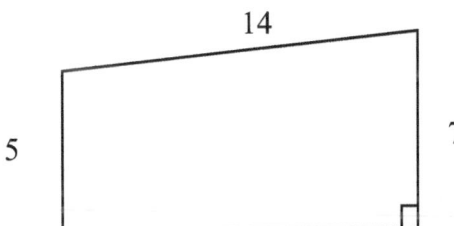

Write your answer in the box below.

Session 2

❖ **Calculators are NOT permitted for this practice test.**

❖ **Time for Session 2: 85 Minutes**

11) A rope weighs 400 grams per meter of length. What is the weight in kilograms of 14.2 meters of this rope? (1 kilograms = 1,000 grams)

    A. 0.0568

    B. 0.568

    C. 5.68

    D. 5,680

12) 5 yards 8 feet and 15 inches equals to how many inches?

    Write your answer in the box below.

13) Which expression has a value of − 19?

    A. 8 − (− 6) + (− 33)

    B. − 3 + (− 7) × (− 4)

    C. − 3 × (− 9) + (− 4) × (− 2)

    D. (− 6) × (− 5) +3

14) Of the 2,600 videos available for rent at a certain video store, 390 are comedies. What percent of the videos are comedies?

    A. $30\frac{1}{4}$%

    B. 25%

    C. 18%

    D. 15%

15) The length of a rectangle is $\frac{2}{7}$ of inches and the width of the rectangle is $\frac{3}{8}$ of inches. What is the area of that rectangle?

A. $\frac{5}{12}$

B. $\frac{1}{28}$

C. $\frac{3}{28}$

D. $\frac{7}{12}$

16) William keeps track of the length of each fish that he catches. Following are the lengths in inches of the fish that he caught one day: 8, 14, 4, 13, 2, 4, 18

What is the median fish length that William caught that day?

A. 2 Inches

B. 13 Inches

C. 4 Inches

D. 8 Inches

17) Solve. $\frac{1}{2} + \frac{7}{8} - \frac{6}{16} =$

A. $\frac{13}{16}$

B. $\frac{9}{16}$

C. 1

D. 0

18) If one acre of forest contains 130 pine trees, how many pine trees are contained in 24 acres?

   A. 480

   B. 6,240

   C. 3,120

   D. 12,480

19) How many $\frac{1}{5}$ cup servings are in a package of cheese that contains $6\frac{1}{5}$ cups altogether?

   A. $7\frac{1}{5}$

   B. $\frac{18}{5}$

   C. 30

   D. 31

20) The area of the base of the following cylinder is 36 square inches and its height is 7 inches. What is the volume of the cylinder?

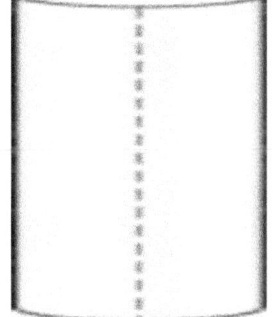

Write your answer in the box below.

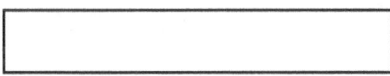

"This is the end of Practice Test 1."

… Georgia Milestones Assessment System

Practice Test 2

Mathematics

GRADE 5

Released

Session 1

- ❖ **Calculators are NOT permitted for this practice test.**
- ❖ **Time for Session 1: 85 Minutes**

GMAS Subject Test Mathematics Grade 5

1) Jack added 16 to the product of 11 and 14. What is this sum?

   A. 154

   B. 720

   C. 170

   D. 980

2) Joe makes $5.12 per hour at his work. If he works 6 hours, how much money will he earn?

   A. $28.72

   B. $38.72

   C. $25.72

   D. $30.72

3) Which of the following is an obtuse angle?

   A. 79°

   B. 45°

   C. 170°

   D. 190°

4) What is the value of $8 - 4\frac{3}{7}$?

   A. $\frac{24}{7}$

   B. $3\frac{4}{7}$

   C. $-2\frac{1}{7}$

   D. $4\frac{4}{7}$

GMAS Subject Test Mathematics Grade 5

5) The bride and groom invited 315 guests for their wedding. 270 guests arrived. What percent of the guest list was not present?

   A. 19%

   B. 18%

   C. 29.56%

   D. 14.28%

6) In a party, 17 soft drinks are required for every 40 guests. If there are 280 guests, how many soft drinks are required?

   A. 11

   B. 91

   C. 119

   D. 158

7) You are asked to chart the temperature during an 8–hour period to give the average. These are your results:

   7 am: 3 degrees   11 am: 20 degrees

   8 am: 5 degrees   12 pm: 30 degrees

   9 am: 11 degrees   1 pm: 31 degrees

   10 am: 19 degrees   2 pm: 33 degrees

   What is the average temperature?

   A. 19

   B. 21.5

   C. 18

   D. 20.25

8) While at work, Emma checks her email once every 40 minutes. In 16 hours, how many times does she check her email?

Write your answer in the box below.

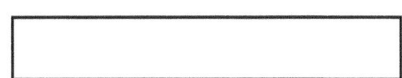

9) A florist has 950 flowers. How many full bouquets of 19 flowers can he make?

Write your answer in the box below.

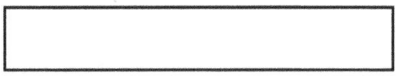

10) What is 5,189.47645 rounded to the nearest tenth?

A. 5,189.476

B. 5,189.5

C. 5,189

D. 5,189.48

Session 2

- ❖ Calculators are NOT permitted for this practice test.
- ❖ Time for Session 2: 85 Minutes

GMAS Subject Test Mathematics Grade 5

11) What is the volume of the following rectangle prism?

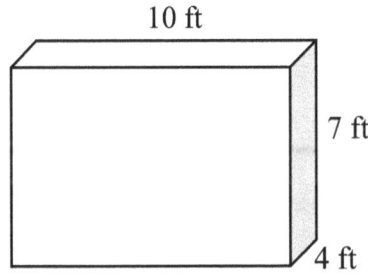

A. $21\ ft^3$

B. $228\ ft^3$

C. $70\ ft^3$

D. $280\ ft^3$

12) A circle has a diameter of 6 inches. What is its approximate circumference?

($\pi = 3.14$)

A. 9.42 inches

B. 18.84 inches

C. 36.54 inches

D. 26.55 inches

13) How long is the line segment shown on the number line below?

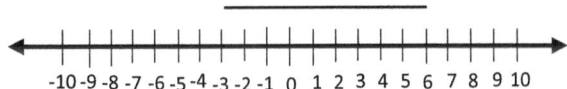

A. 9

B. 11

C. 3

D. 7

GMAS Subject Test Mathematics Grade 5

14) Peter traveled 180 miles in 9 hours and Jason traveled 700 miles in 10 hours.

    What is the ratio of the average speed of Peter to average speed of Jason?

    A. 2: 7

    B. 4: 7

    C. 7: 5

    D. 7: 3

15) If $x = -5$, which equation is true?

    A. $x(3x + 7) = 45$

    B. $4(6 - x) = 44$

    C. $3(2x + 13) = 12$

    D. $3x - 16 = -34$

16) What are the coordinates of the intersection of x-axis and the y-axis on a coordinate plane?

    A. $(-1, -1)$

    B. $(7, 0)$

    C. $(0, 0)$

    D. $(0, 7)$

17) In a triangle ABC the measure of angle ACB is 48° and the measure of angle CAB is 63°. What is the measure of angle ABC?

    Write your answer in the box below.

18) Which list shows the fractions listed in order from least to greatest?

$$\frac{1}{11}, \frac{1}{15}, \frac{1}{4}, \frac{1}{9}$$

A. $\frac{1}{11}, \frac{1}{4}, \frac{1}{15}, \frac{1}{9}$

B. $\frac{1}{9}, \frac{1}{15}, \frac{1}{4}, \frac{1}{11}$

C. $\frac{1}{4}, \frac{1}{11}, \frac{1}{9}, \frac{1}{15}$

D. $\frac{1}{15}, \frac{1}{11}, \frac{1}{9}, \frac{1}{4}$

19) If a rectangular swimming pool has a perimeter of 88 feet and it is 18 feet wide, what is its area?

Write your answer in the box below.

20) Aria was hired to teach six identical 3rd grade math courses, which entailed being present in the classroom 36 hours altogether. At $29 per class hour, how much did Aria earn for teaching one course?

A. $130

B. $174

C. $810

D. $1,320

"This is the end of Practice Test 2."

Chapter 14 : Answers and Explanations

Georgia Milestones Assessment System
Answer Key

❋ Now, it's time to review your results to see where you went wrong and what areas you need to improve!

Practice Test - 1

1	B	11	C
2	A	12	291
3	D	13	A
4	A	14	D
5	B	15	B
6	16π	16	D
7	D	17	C
8	D	18	C
9	A	19	D
10	72	20	252

Practice Test - 2

1	C	11	D
2	D	12	B
3	C	13	A
4	B	14	A
5	D	15	B
6	C	16	C
7	A	17	69°
8	24	18	D
9	50	19	468
10	B	20	B

GMAS Subject Test Mathematics Grade 5

Practice Test 1
Georgia Milestones Assessment
System - Mathematics
Answers and Explanations

1) Answer: B.

6 eggs for 1 cake. Therefore, 240 eggs can be used for (240 ÷ 6) 40 cakes.

2) Answer: A.

Use area of rectangle formula.

$area\ of\ a\ rectangle\ =\ width\ \times\ length \Rightarrow D = w \times l \Rightarrow w = \frac{D}{l} = \frac{D}{9}$

3) Answer: D.

To list the fractions from least to greatest, you can convert the fractions to decimal.

$\frac{5}{8} = 0.625; \frac{4}{9} = 0.44; \frac{8}{10} = 0.8; \frac{10}{11} = 0.909; \frac{3}{13} = 0.272$

$\frac{3}{13} = 0.272, \frac{4}{9} = 0.44, \frac{5}{8} = 0.625, \frac{8}{10} = 0.8, \frac{10}{11} = 0.909$

Option D shows the fractions in order from least to greatest.

4) Answer: A.

Plug in 25 for A in the equations. Only option A works.

$A + 25 = 50 \Rightarrow 25 + 25 = 50$

5) Answer: B.

4 multiplied by $\frac{6}{5} = \frac{24}{5} = 4.8$, therefore, only choice B is correct.

6) Answer: 16π.

Use area and circumference of circle formula.

Area of a circle = $\pi r^2 \Rightarrow 64\pi = \pi r^2 \Rightarrow r = 8$

Circumference of a circle = $2\pi r \Rightarrow C = 2 \times 8 \times \pi \Rightarrow C = 16\pi$

7) Answer: D.

Use volume of rectangle formula.

$Volume\ of\ a\ rectangle\ =\ width\ \times\ length\ \times\ heigth \Rightarrow V = 4 \times 3 \times 7 \Rightarrow V = 84$

WWW.MathNotion.Com

GMAS Subject Test Mathematics Grade 5

8) Answer: D.

To find the selling price, multiply the price by (100% – rate of discount).

Then: (180) (100% – 30%) = (180) (0.70) = 126

9) Answer: A.

There are 50 cards in the bag and 4 of them are white. Then, 4 out of 50 cards are white. You can write this as: $\frac{4}{50}$. To simplify this fraction, divide both numerator and denominator by 2. Then: $\frac{4}{50} = \frac{2}{25}$

10) Answer: 72.

First, find the missing side of the trapezoid. The perimeter of the trapezoid below is 38. Therefore, the missing side of the trapezoid (its height) is:

$38 - 5 - 7 - 14 = 38 - 26 = 12$

Area of a trapezoid: $A = \frac{1}{2} h (b1 + b2)$

$= \frac{1}{2}(12)(5 + 7) = 72$

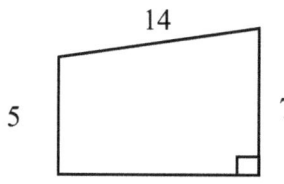

11) Answer: C.

1 meter of the rope = 400 grams

14.2 meter of the rope = 14.2 × 400 = 5,680 grams = 5.68 kg

12) Answer: 291.

5 yards = 5 × 36 = 180 inches

8 feet = 8 × 12 = 96 inches

5 yards 8 feet and 15 inches = 180 inches + 96 inches + 15 inches = 291 inches

13) Answer: A.

Simplify each option provided using order of operations rules.

A. 8 – (– 6) + (– 33) = 8 + 6 – 33 = –19

B. – 3 + (– 7) × (– 4) = –3 + 28 = 25

C. – 3 × (– 9) + (– 4) × (– 2) = 27 + 8 = 35

D. (– 6) × (– 5) + 3 = 30 + 3 = 33

Only option A is −19.

GMAS Subject Test Mathematics Grade 5

14) Answer: D.

Use percent formula: part = $\frac{percent}{100} \times$ whole

$390 = \frac{percent}{100} \times 2{,}600 \Rightarrow 390 =$ percent $\times 26 \Rightarrow$ percent $= 15$

15) Answer: B.

Use area of rectangle formula.

$Area = length \times width \Rightarrow A = \frac{2}{7} \times \frac{3}{8} \Rightarrow A = \frac{3}{28}$ inches

16) Answer: D.

Write the numbers in order: 2, 4, 4, 8, 13, 14, 18.

Median is the number in the middle. Therefore, the median is 8.

17) Answer: C.

$\frac{1}{2} + \frac{7}{8} - \frac{6}{16} = \frac{(8 \times 1)+(2 \times 7)-(6 \times 1)}{16} = \frac{16}{16} = 1$

18) Answer: C.

1 acre: 130 pine trees

24 acres: $130 \times 24 = 3{,}120$ pine trees

19) Answer: D.

To solve this problem, divide $6\frac{1}{5}$ by $\frac{1}{5}$.

$6\frac{1}{5} \div \frac{1}{5} = \frac{31}{5} \div \frac{1}{5} = \frac{31}{5} \times \frac{5}{1} = 31$

20) Answer: 252.

Use volume of cylinder formula.

$Voluem = base \times heigth \Rightarrow V = 36 \times 7 \Rightarrow V = 252$

Practice Test 2
Georgia Milestones Assessment
System - Mathematics
Answers and Explanations

1) Answer: C.

$16 + (11 \times 14) = 16 + 154 = 170$

2) Answer: D.

1 hour: $5.12; 6 hours: $6 \times \$5.12 = \$30.72$

3) Answer: C.

An obtuse angle is an angle of greater than 90° and less than 180°. From the options provided, only option C (170 degrees) is an obtuse angle.

4) Answer: B.

$8 - 4\frac{3}{7} = \frac{56}{7} - \frac{31}{7} = \frac{25}{7} = 3\frac{4}{7}$

5) Answer: D.

The number of guests that are not present are: (315 − 270) 45 out of 315 = $\frac{45}{315}$

Change the fraction to percent: $\frac{45}{315} \times 100\% = 14.28\%$

6) Answer: C.

Write a proportion and solve.

$\frac{17 \text{ soft drinks}}{40 \text{ guests}} = \frac{x}{280 \text{ guests}} \Rightarrow x = \frac{280 \times 17}{40} \Rightarrow x = 119$

7) Answer: A.

average (mean) = $\frac{\text{sum of terms}}{\text{number of terms}} \Rightarrow$ average = $\frac{3+5+11+19+20+30+31+33}{8} \Rightarrow$ average = 19

8) Answer: 24.

Every 40 minutes Emma checks her email.

In 16 hours (960 minutes), Emma checks her email (960 ÷ 40) 24 times.

9) Answer: 50.

Divide the number flowers by 19: 950 ÷ 19 = 50

GMAS Subject Test Mathematics Grade 5

10) Answer: B.

Rounding decimals is similar to rounding other numbers. If the hundredths and thousandths places of a decimal is forty-seven or less, they are dropped, and the tenths place does not change. For example, rounding 0.923 to the nearest tenth would give 0.9. Therefore, 5,189.47645 rounded to the nearest tenth is 5,189.5.

11) Answer: D.

Use volume of rectangle prism formula.

$V = length \times width \times height \Rightarrow V = 4 \times 7 \times 10 \Rightarrow V = 280$

12) Answer: B.

The diameter of the circle is 6 inches. Therefore, the radius of the circle is 3 inches.
Use circumference of circle formula: $C = 2\pi r \Rightarrow C = 2 \times 3.14 \times 3 \Rightarrow C = 18.84$

13) Answer: A.

The line segment is from 6 to -3. Therefore, the line is 9 units.

$6 - (-3) = 6 + 3 = 9$

14) Answer: A.

Peter's speed $= \frac{180}{9} = 20$; Jason's speed $= \frac{700}{10} = 70$

$\frac{The\ average\ speed\ of\ peter}{The\ average\ speed\ of\ Jason} = \frac{20}{70}$ equals to: $\frac{2}{7}$ or 2: 7

15) Answer: B.

Plug in $x = -5$ in each equation.

A. $x(3x + 7) = 45 \rightarrow (-5)(3(-5) + 7) = (-5) \times (-15 + 7) = 40$

B. $4(6 - x) = 44 \rightarrow 4(6 - (-5)) = 4(11) = 44$

C. $3(2x + 13) = 12 \rightarrow 3(2(-5) + 13) = 3(-10 + 13) = 9$

D. $3x - 16 = -34 \rightarrow 3(-5) - 16 = -15 - 16 = -31$

Only option B.

16) Answer: C.

The horizontal axis in the coordinate plane is called the $x - axis$. The vertical axis is called the $y - axis$. The point at which the two axes intersect is called the origin. The origin is at 0 on the $x - axis$ and 0 on the $y - axis$.

GMAS Subject Test Mathematics Grade 5

17) Answer: 69°.

All angles in every triangle add up to 180°. Let x be the angle ABC.

Then: $180 = 48° + 63° + x \Rightarrow x = 69°$

18) Answer: D.

In fractions, when denominators increase, the value of fractions decrease and as much as numerators increase, the value of fractions increase. Therefore, the least one of this list is: $\frac{1}{15}$ and the greatest one of this list is: $\frac{1}{4}$

19) Answer: 468.

Perimeter of rectangle formula:

$P = 2\ (length\ +\ width) \Rightarrow 88 = 2\ (l + 18) \Rightarrow l = 26$

Area of rectangle formula: $A = length \times width \Rightarrow A = 26 \times 18 \Rightarrow A = 468$

20) Answer: B.

Aria teaches 36 hours for six identical courses. Therefore, she teaches 6 hours for each course. Aria earns $29 per hour. Therefore, she earned $174 (6 × 29) for each course.

"End"

www.ingramcontent.com/pod-product-compliance
Lightning Source LLC
Chambersburg PA
CBHW080441110426
42743CB00016B/3235